Combat Aircraft of World War II

TWO VOLUMES IN ONE

Glenn B. Bavousett

BONANZA BOOKS / New York

This edition contains the complete and unabridged texts of the editions. They have been completely reset for this volume.

Originally published in separate volumes under the titles:
World War II Aircraft in Combat, copyright © 1976, 1974 by Glenn B. Bavousett
More World War II Aircraft in Combat, copyright © 1981 by Glenn B. Bavousett
All rights reserved.

This 1989 edition is published by Bonanza Books, distributed by Crown Publishers, Inc., 225 Park Avenue South, New York, New York 10003, by arrangement with the author.

Printed and Bound in Singapore

Library of Congress Cataloging-in-Publication Data

Bavousett, Glenn B.
   Combat aircraft of World War II/Glenn B. Bavousett.
      p.   cm.
   ISBN 0-517-68019-X
   1. Airplanes, Military—Pictorial works.   2. World War, 1939–1945—
Aerial operations—Pictorial works.   I. Title.
UG1240.B38   1989
358.4′183′0222—dc20   89-7307
                                                              CIP

h g f e d c b a

Contents

Publisher's Note

In order to combine Glenn Bavousett's two magnificent books into one comprehensive volume, we found it necessary to place the color plates from both works together in one section at the back of the new edition. From time to time, the descriptive text for a plane refers to a detail or scene depicted in the corresponding painting. The original wording has been left intact, as we feel that reading the histories of these fascinating aircraft only whets the appetite further for the extraordinary paintings to come.

VOLUME I

Acknowledgments

This book would have been impossible without considerable encouragement and assistance from various quarters. The generosity of many individuals and organizations who furnished information is deeply appreciated.

Special appreciation is accorded to William N. Hess, author, air historian and secretary of the AMERICAN FIGHTER ACES ASSOCIATION, who continually encouraged me and furnished meaningful data; Christopher Shores of London, England, for the relentless critiques that were solicited of him; Bernard Millot of Montrueil, France, who bailed me out on occasion and plugged some holes; and Gene Stafford, author, air historian and illustrator who provided critical information when it was needed. All of these talented men made significant contributions to what is read and seen.

It was an exciting experience performing the research necessary to find the stories and develop the resulting scenes. In the process many sources were tapped for information or critique. Some may have escaped getting into my files, and to those of you whom I may have missed in the listing below, please accept my apologies.

American Fighter Aces Association; Bell Aerospace Company; Association Française pour l'Histoire de l'Aviation; Service Historique de l'Armée de l'Air; Forces Aeriennes Françaises Libres; 98th Bomb Group Veterans Association (The Pyramidiers); The Albert F. Simpson Historical Research Center; American Aviation Historical Society; former members of the AVG; National Archives; 20th Air Force Association; Naval Air Museum; The Boeing Company; and the Confederate Air Force.

Denny D. Pidhayny; LCDR Tom Legget, Jr; Capt. Grover Walker, USN; Lee Pearson, Navy Department; Arthur L. Schoeni; Col. F.C. Caldwell, USMC; Alain Romans; Royden LeBrecht; John A. Fornwalt; Col. Ulysses S. Nero; Dr. Hal Fenner; Col. John R. Kane; Eddie Holland; Charles G. Worman; Gen. Joe Kilgore; Col. Lloyd P. Nolen, CAF; Roger Freeman; Len Morgan; Col. Kenneth H. Dahlberg; James J. Sloan; Col. Hubert "Hub" Zemke; Robert T. Smith; David Lee "Tex" Hill; Gen. C.T. "Curly" Edwinson; William E. Blurock; Lee "Kitty" Carr; Gale Hasenplaugh; Jean Coleman; Daniel T. Goggin; Leslie Waffen; Richard Keenan; Col. James Patillo; Joseph Pokraka; Col. Eino Jenstrom; Carter McGregor; Victor Agather; Gen William Skaer; Col. T. M. D'Andrea, USMC; Charles W. Cain; Bruce Bennett; Hans-Joachim Kroschinski; J. Frank Dial, and Rudolph Schmeichel.

3

Foreword

Glenn Bavousett and Tony Weddel have done a magnificent job of joining some of the finest, most realistic military aviation paintings I have ever seen to exciting but concise explanatory text.

While this volume is not a detailed history of the air action in the Second World War, it most certainly does depict, in true color, the principal aircraft that took part in the conflict. Here are the *Mustangs, Thunderbolts, Spitfires, Hurricanes,* Bf 109s, *Yaks, Zeros, Wildcats, Hellcats, Corsairs, Kingfishers, Stukas, Mosquitos, Helldivers, Vals, Mitchells, Liberators, Flying Forts,* and *Superforts* . . . and more.

They were the fighters, scouts, liaison planes, patrol bombers, attack aircraft, torpedo and dive bombers and medium and heavy bombers that fought for the mastery of the sky and the domination of the earth below. And many of those who flew them gave their lives for it.

Along with the principal military and naval aircraft illustrated here are such other fighter planes as the French flown Curtiss 75A *Hawk,* the German FW-190, and the jet Me 262; bombers, including the German Heinkel He 111 and Japanese *Betty;* and the massive 6-engine cargo aircraft, the Me 323 *Gigant.*

At first glance, and even at second glance, many of the paintings appear to be highly detailed full-color photos taken in the midst of battle—where, in fact, no camera had ventured. Actually, one might say that the artist's brush captured on canvas what the camera's lens failed to put on film.

Devotion to accuracy and detail necessitated the reconstruction of each battle, using scale models. Facts were gathered from available official records and by means of personal interviews. Some of these interviews took the author many thousands of miles. Meetings with pilots or other participants in the action sometimes resulted in new information which necessitated redoing entire paintings.

This is a fine book and the creative team that put it together are to be commended for a job well done!

STANLEY M. ULANOFF

Introduction

The following pages present a graphic, historical account of World War II aerial combat. Out of necessity the presentation is limited in scope. What you will see and read has taken years to research, write and illustrate. Aerial combat during World War II is so story rich that it would be virtually impossible to conclude a highly comprehensive graphic presentation.

Aircraft types and the actions in which they appear were selected to provide a loosely connected chain of events that allows the reader to trace the progression of the war, yet leaves gaps to be filled through the adventure of gaining additional information from other sources.

The U.S.A.F. is represented in thirteen of the paintings. The U.S. Navy and Marine Corps are covered in ten, and foreign markings, on both friend and foe, are shown in thirteen more. The original canvases are in three sizes: 24 × 32 inches was used for single engine aircraft, 32 × 40 for multi-engine and one 40 × 48 which was required for the Missing Man Formation.

With the exception of a couple of scenes you will note an absence of some well-known actions, and in a few instances less prominent, yet interesting, aircraft are depicted. For example, no attempt was made to portray the Battle of Britain, Pearl Harbor action or the Doolittle Raid on Japan. These battles are generally well known and are well documented in many writings.

Attention should be paid to both the secondary aircraft depicted and when applicable the background itself. By doing so you will discover not thirty-five different aircraft types but forty-three. Also, you will find landmarks familiar to pilots and crew members.

It should be mentioned that regardless of the amount of effort expended in research to produce as accurate an accounting as possible, error may have found its way into the book. Some paintings resulted from eyewitness accounts. In some instances more than one eyewitness was interviewed and different points of view of the same action were gained. In fast-breaking combat, who stops to take a picture or even make it a point to fix everyone's position? Also, it was found that in some instances even the official record managed to conflict with itself. This in no way is intended to be a qualifying remark or to cover any error of omission or commission. As a matter of record, substantiated constructive criticism is solicited.

This book was developed by a team. About half of the scenes were completely portrayed by the finish artist, Tony Weddel. The others were developed first on vellum stock then transferred onto the canvas by the detail artist, Donovan Gatewood. Professor Neil Duncan of Texas Wesleyan College fulfilled the role of grammarian and standardized procedures regarding certain aspects of the story-lines. Editing was done by Dr. Stan Ulanoff, also a college professor and author/editor of eighteen military aviation books.

Finally, the Flying Tigers did not use the *Warhawk* variant of the

P-40 as has been stated herein, their model was named the *Tomahawk.* *Warhawk* has been used for the simple reason that it has been best remembered under that name. Also, we refer to the Bell *Airacobra* as the P-39 when in fact the aircraft depicted was the export version which was designated P-400.

All of the primary aircraft types presented here existed in collections or museums at the time of this writing. Not all are to be found under one roof. Various museums around the country will produce all of these plus many more.

It is hoped that this book is as interesting and thought provoking to you as it has been to those of us that have been an integral part of its development.

GLENN B. BAVOUSETT

Prologue

Since the late 1930s tremendous strides have been made in the advancement-of-the-state-of-the-art relating to all aspects of aviation. The "great" leap certainly occurred during the war years from 1939 through 1945. The pressure of the times served as an impetus to the aviation industry. In this period hundreds of thousands of aircraft were designed and built. Quite a few were either one-of-a-kind experimental types or were produced in very limited quantity, for one reason or another. Out of it all, however, emerged a surprisingly large number of types which were subsequently produced in great quantities.

It was a time of unparalleled ingenuity. The German Luftwaffe was not only sophisticated when stacked up against any other Air Force, it also counted heavily in numbers of aircraft and experienced airmen. Japan was no pushover either when it came to aerial might. Being a nation of islands it was only logical that they would have a vast naval armada manned by excellent seamen. Unlike Germany, Japan's navy included first class carriers as an integral part of her fleet. As a result the Imperial Japanese Navy could boast of having not only the very latest aerial weaponry but also tough, well-disciplined first line pilots and crews. Their effectiveness was keenly displayed not only at Pearl Harbor but in other battles throughout the Far East and the Pacific.

Germany stunned the world with the preemptive *blitzkrieg.* Always present was the invincible Luftwaffe blasting the pathways for Hitler's legions of tanks and fast moving troops.

No nation was prepared for either the German or Japanese onslaughts. As a result all resistance in Europe ended temporarily on blood-soaked beaches at Dieppe. The road back would have to begin in England. The situation in the Pacific was equally bleak. Pockets of resistance held out unbelievably long. In the end, they, too, fell beneath the heavy boot of the Japanese soldier. Precious little of the Pacific and Asia remained free of Japanese occupation; that which did was permitted to be so not because of defensive strength but because the Japanese elected to "do business" elsewhere. The road back in Asia began in several places dictated mainly by geography; Hawaii, Australia and Burma-India, primarily.

In all of the theaters of war, however, one common objective was mandatory; to achieve victory you had to have aerial superiority—and that went for both sides. Because of that, legends of aerial combat became abundant. In fact historians are still digging into the seemingly endless stack of undocumented stories.

World War II presented a forum for innovativeness in aerial combat, both strategic and tactical. If necessity is the mother of invention, then this great conflict opened the flood gates, wide. In a short span of time pilots were transitioning from obsolete, undergunned slow moving planes to power-laden aircraft complete with excellent radar and racks of missiles slung beneath the wings. Entering the arena were also such aerial weapons as the V-1 and V-2 rockets, jet and rocket powered aircraft, suicide aircraft and piloted bombs. It seemed nothing was impossible and every day something new was tried. Pure muscle, resolution and massive supplies of material pitted against him ended it for Hitler in Europe; a pair of bombs with the greatest punch the world had ever known helped the Japanese to an early way out in the Pacific.

This was the beginning of the end for the human element playing the dominant role in aerial combat. The lessons learned by industry would soon have the speeds exceeding not only the speed of sound but doubling and tripling it. No longer would we need hundreds of thousands of aircraft—a literal handful would be all that would be necessary in future contests. From now on they would be weapon platforms guided by electronics and computers . . . the human factor in combat would recede rapidly.

What all this means is that personal aerial combat between men and machines peaked out during World War II. The days of the Aces and 30, 40, 150 or 200 "kills" are gone forever. And with them went the tales of aerial combat.

FIGHTERS

Curtiss P-40 *Warhawk (Tomahawk)*

Fei Weing, Flying Tigers, was the name given by the Chinese to the rag-tag unconventional pilots of the fighting A. V. G. (American Volunteer Group) who stalled the Japanese advance into Burma, then saved China in a brief series of heroic battles at the Salween River near the Kunming, China, the end of the famous Burma Road.

It cannot be said that America does not have its hawks, men born with a lust for combat. Without our hawks and highly innovative leaders such as Claire Lee Chennault, our victories on land, sea, and in the air might never have been.

Men of vision, such as Chennault, knew this country was destined for war. In the case of Chennault he knew all of China, Burma, India, and Southeast Asia would fall to the Japanese unless an assemblage of highly dedicated American military pilots were not formed in Asia well in advance of our forced entry into that war. Moreover, they would require total support from home including the late model P-40s.

His bluntness and reputation of being a *prima donna* left Chennault with few friends in the military, but his overpowering, accurate assessment of the Japanese Air Force capabilities and limitation gained him the ear of President Roosevelt, who in turn set in motion the directives that would give "Old Leatherface" Chennault his A. V. G.

It was miserably slow in coming, but when it did Chennault was able to put together three of the meanest Fighter Squadrons ever assembled on the face of the earth. The Hell's Angels operated initially out of the Rangoon, Burma area, while the Adams and Eves and Panda Bears were at Kunming, the other end of the narrow switchbacking Burma Road.

Chennault's first task with his one hundred or so ex-Army, Navy, and Marine pilots, who had been allowed to resign from service to sign up for a year's duty with the AVG, was to untrain them. The usual conventional fighter tactics made the P-40 inferior to the lighter more highly maneuverable enemy aircraft. To effectively defeat them meant all fighting would have to be done employing the strong points of the P-40. This meant attacking in pairs rather than singly or in groups, diving down into the highly disciplined Jap formations that would not break and run, or barrel their P-40s head-on into them but never be lured into dogfights or turn or climb. Here the P-40 was extremely vulnerable. These Chennault-devised tactics violated all rules of combat imbedded in his mixed fighting force. But, years of actual battle against these Japanese planes by Chennault himself (seventy-five downed from the cockpit of antiquated equipment) proved conclusively that his radical offensive tactics were proper and deadly effective.

Casting aside all vestiges of military bearing, the men of the AVG maintained whatever lifestyle they preferred. None wore standard uniforms, not even Chennault. Shorts, loincloths, cowboy boots, and hip-slung six shooters were not at all uncommon sights among the pilots and their ground crews. Chennault was bent on results, not tidiness or decorum.

Chennault kept his AVG mobile, constantly moving them to hurriedly constructed air strips from which to operate and confuse the orderly Japanese mind into believing far more aircraft existed than the actual one hundred.

These unconventional strategies and fighting tactics were never solved or effectively countered by the enemy. Their only answer was to pour in increasingly more aircraft which only resulted in a more staggeringly high number of enemy aircraft destroyed by the AVG pilots.

This scene captures a pair of P-40s clearing a piece of Burma sky of Mitsubishi *Claudes,* several of some 286 enemy aircraft scorched from the skies in one eight-month period of action. In the foreground is a Hell's Angels Squadron plane as evidenced by the little red lady just beneath and forward of the cockpit on the fuselage. Above is a member of the Panda Bears. It was not at all uncommon for elements of all three Squadrons to fly together.

The combined record of the AVG, 10th and 14th Air Forces is unknown; however, in excess of one thousand enemy aircraft were destroyed in the sky and another three thousand on the ground. Sunk or damaged shipping certainly ran over 2,500,000 tons while absolutely no meaningful estimate is available on the ground damage sustained in the countless strafing/bombing attacks.

Contrary to popular opinion, the sharkmouths were not spawned by the AVG. They got the idea from a magazine article showing British P-40 fighter activity in North Africa.

While the best known model of the P-40 was known as the *Warhawk*, the specie flown by the AVG was called the *Tomahawk*. Others were named *Kittyhawk*.

Hawker *Hurricane*

The classy *Spitfire* is remembered as the machine that fought the Battle of Britain, but in reality it was the rugged *Hurricane* that fought the hard fight.

It was *Hurricanes* that were sent to France to do battle with the much vaunted Luftwaffe. They were there all during the Phony War roaring into the skies every day, long before Dunkirk and the Battle of Britain; the *Spitfire,* the *Hurricane's* stablemate, had yet to arrive on the battle scene. These *Hurricane* pilots went up against the mighty Luftwaffe on a daily basis and as a result they gained the tactical experience they passed on to the pilots who would fight in the Battle of Britain.

Both the *Hurricanes* and *Spitfires* gave good accounts of themselves against the Germans during the evacuation at Dunkirk. By the beginning of July, 1940, the opening of the Battle of Britain, the Royal Air Force Fighter Command had 44 squadrons, 25 of which were *Hurricanes.* During fighting in France it was found that the *Hurricane* was out-performed by the sleek Messerschmitt Bf 109; therefore, whenever possible in the Battle of Britain fighting a squadron of *Spitfires* went into action along with a squadron of *Hurricanes.* The "Spits" would take on the escorting Messerschmitts while the gun-laden *Hurricanes* plowed into the bomber formations. This team arrangement proved to be highly successful in minimizing losses while tearing up the enemy.

The Battle of Britain produced many stories of high adventure, courage and bravery, and brought the names of numerous R.A.F. pilots to the fore. One of these was Sergeant Pilot J.H. "Ginger" Lacey, a red-haired *Hurricane* pilot who bagged 28 enemy aircraft, added four probables and damaged eleven more. "Ginger" was unique in many ways; he was on the line ready to fight when the war broke out and he was still

airborne and fighting when the war ended. He fought both in Europe and the Far East. Lacey's last "kill" was a Japanese *Oscar* that he got on the 19th of February in 1945, a long time after his victories in the Battle of Britain. "Ginger" was noted for being extremely cool in combat and his willingness to get into the thick of things. His fighting spirit was well demonstrated in the nine times he was shot down or crash landed his *Hurricanes.* Lacey bailed out more than once.

One of Lacey's better days was 15 September 1940 during the Battle of Britain. On this day he rose to face the invading Luftwaffe two times and he scored four victories. One was the victory over a Heinkel 111 twin-engined bomber, depicted here.

Supermarine *Spitfire*

On October 12, 1940, the Battle of Britain was over. Adolph Hitler had elected to postpone *Operation Sealion,* the code name given for the invasion of the British Isles, and shifted his attention to *Operation Barbarossa,* the master plan for the invasion of Russia.

Included in the master plan of *Sealion* was the movement of large forces and supplies in huge gliders designed for the short one-way trip across the channel. Both Messerschmitt and Junkers went into production of these great gliders. With the end of *Sealion* came the end of the glider program, and a lucky fate for those German soldiers and airmen who would have been in them. Without question, the legendary *Spitfires* and their fearless pilots would have torn them to ribbons.

Between the fall of 1940 in Britain and the spring of 1943 in North Africa, the great Messerschmitt glider had undergone a remarkable transformation from no engines to four engines and then to six.

In that same period of time the classic "Spit" had also undergone a few changes.

Tunisia would be the battleground for these two aircraft rather than the English Channel and British Isles. A steady flow of men and materials were being shuttled to Tunis from Luftwaffe bases in Italy and Sicily. Junkers Ju52 transports and Messerschmitt Me323 *Gigants* with Bf 109s and 110s flying escort made the low-level runs practically unmolested during the latter months of 1942.

In late March of 1943 the Battle of Mareth, Tunisia, was begun and the line was outflanked. With Rommel ill and out of North Africa the Axis defenses began to crumble under the unrelenting pressure being applied by the British and Americans. It was during this stage of the collapse of the German presence in North Africa that *Operation Flax* was launched by the Allies. *Flax* was an intensified attempt to interrupt the Axis air transport system linking Europe and Africa.

By springtime the Allied air power in the area was such that 18 April 1943 would go down in the annals of aerial warfare as "Palm Sunday Massacre." On this day a large flight of German transports and covering fighters were caught by 9th Air Force P-40s and R.A.F. *Spitfires*. In this one brief action 59 of the transports were claimed and the top covering *Spitfires* accounted for sixteen of the enemy fighters.

Four days later, 22 April, saw the slaughter of the Messerschmitt Me323 *Gigants*. Sixteen of the enormous brutes, laden with badly needed fuel, came low across the Mediterranean—high above them flew their Bf 109 and 110 escorts. In the air for the Allies were several flights of fighters, predominately South African Air Force aircraft. Flying top cover was a Polish Fighting Team (145 R.A.F. Squadron) in Mk IX *Spitfires* plus 1 S.A.A.F. Squadron, a 244 Wing unit, with *Spitfire* Vs, while the main attack force was 7 S.A.A.F. Wing *Kittyhawks* and an R.A.F. *Kittyhawk* unit. The clash came off Cap Bon, Tunisia, and in the fierce battle all but one of the *Gigants* were destroyed by the South Africans. Practically all of the 323s fell to the guns of the *Kittyhawks,* however, "Spit" Vs from a S.A.A.F. Squadron claimed five of the big transports. The top covering Poles stayed clear of the fracas below them and held the German fighters at bay while the South Africans pulverized the 323s.

In this scene you see a Mk IX *Spitfire* piloted by the Polish ace Flight Lieutenant Eugeniusz "Horba" Horbaczewski, who was the Poles' top scorer in Africa. In all "Horba" chalked up 16½ victories in both Africa and Europe; this tally does not count his V-1 flying bombs shot down. On 18 August 1944 "Horba" shot down three FW 190s but lost his life in this combat. In the background of the painting a doomed *Gigant* is in its plunge toward the Mediterranean.

Without question the classic *Spitfire* will have its special place in our air history as being one of the few truly superb fighter aircraft types of all times.

Republic P-47 *Thunderbolt* (*Razorback*)

Colonel Hubert "Hub" Zemke was an outstanding leader, aggressive combat pilot and an excellent tactician. He was boss of the 56th Fighter Group when it left the States and was first to receive the new P-47 Razorback *Thunderbolts*. The Group left for England in early January 1943, but it was not until April that it entered combat for the first time. Zemke brought a knowledge of the enemy the men would face in combat gleaned from service as an observer during the Battle of Britain and later as a P-40 instructor in the Soviet Union. "Hub" scored for the first time on 13 June 1943 when he downed a brace of FW-190s. He added to this score in August and September and became an Ace on 2 October. Shortly after his sixth victory he was told he would have to return to the States to be part of a Goodwill Tour. On the day he was supposed to leave England Zemke showed up ready for flying instead and led his 56th on a B-24 escort mission to Germany. On the mission he collected his seventh victory. Following is his Personal Combat Report filed after returning to base. The action took place on 5 November 1943 at approximately 1350 hours between Enschede, Holland, and Rheine, Germany, at 28,000 feet altitude. It was a clear day and unlimited with lower haze; visibility was good at altitude. This is the Report:

"As Group Commander I was leading the 63rd Fighter Squadron. Just before the combat, the 63rd Fighter Squadron was escorting the B-24's of the 5th Task Force, as given in F.O. 170 of that date. The position of the 63rd Squadron was on the right flank of the lead box of bombers, while the 61st was on the left flank. The bomber formation was proceeding toward Munster, Germany. For some time we had been on escort with no enemy in the vicinity, when I was told there were several FW 190s coming into us from eleven o'clock (from the N.E.). These enemy aircraft were immediately picked up as being ahead by four or five miles and at our same altitude. They were estimated as about thirty staggered in depth. Since the 61st was somewhat behind on the left flank, I ordered the 61st Squadron to cross in front of the bombers and break up the enemy aircraft. We proceeded across the path of the bombers just as the enemy aircraft turned ahead to fly along the path of the bombers. The 61st caught up with us just as the 63rd was converging on the rear of the enemy formation. The enemy nosed down a bit by then, so I stepped up the R.P.M. to 2700, and the manifold to 40″ Hg. At a gradual rate, I closed up on a low FW 190 carrying a belly tank and two rocket guns. At 500 yards the enemy formation was still intact and I was afraid they sensed our presence, so I opened up with a burst of perhaps twenty-five rounds. No strikes were seen to register, so I withheld my fire for a time, still closing in on the FW 190. At 400 yards, I opened again and from that time until I closed within 100 yards, short bursts were fired. No telling effect was registered save for some occasional hits on the wings and fuselage, until at very close range I saw my tracers just going below the Focke-Wulf. The sights were raised somewhat, and the next burst blew the canopy and many pieces from the aircraft. He thereupon nosed over and went straight down as I broke to avert colliding. The other enemy aircraft had by then either broken for the deck or wheeled around to become engaged by the pilots following me. My recovery to the left and into the sun, brought me directly above the ensuing battle, where FW 190s were dog fighting with P-47s. I acted as high cover, to dive on any FW 190 who was gaining on a P-47. During this melee at least three other enemy aircraft were seen to be shot down. As soon as the engagement was ended and the enemy dispersed, the Group was ordered to return to base and I moved out."

Zemke then went on the Goodwill Tour and returned to command the 56th on 19 January 1944. He scored the first ground victory for the 56th on 11 February. His victories in the air continued to mount. By the time he left the unit in August 1944, his total stood at 15¼ victories in the air. He took command of the 479th Fighter Group where he flew P-38s and later, P-51s. He scored another 2.5 victories with this unit and ended the war with 17.75 in the air and 8.5 on the ground.

On 30 October 1944, "Hub" was leading a mission of the 479th, something he had been ordered to stop doing since he was in for promotion to Brigadier General. But he went anyway. The flight of *Mustangs* ran into bad weather and Zemke went down. In the end it was the weather and not the Luftwaffe that proved to be Zemke's greatest enemy. "Hub" was taken prisoner and interrogated by Hans Scharff, Luftwaffe interrogater for downed Allied fighter pilots. Zemke made an entry in Scharff's "Guest Book" which rather sums up his feelings: "It was my wish to stay on the other side of the fence for a time more to run up a few more battles with a classic opponent but fate did me otherwise."

Zemke was a great teacher of aerial warfare. The 56th Fighter Group pilots racked up more than 1,000 victories, 750 of which were in the air, and the Group spawned about 40 Aces.

Messerschmitt Bf 109

The "Star of Africa," Hauptmann Hans Joachim Marseille, is shown here as he scores one of his many victories over the Libyan desert. The aircraft is a Bf 109F-4. On the losing end, plunging earthward toward the hot desert sand dunes below, is a P-40 *Kittyhawk*.

Flamboyant and forever dangerous in the sky, Marseille is reputed to have run up a fantastic score of enemy aircraft destroyed—158! Of this total he is reported to have gained seventeen during three sorties in one day over North Africa. It was not that the P-40 or its pilots were that inferior; this young pilot was gifted with the ability to make lighting fast judgments as to when to fire.

Marseille died on 30 September 1942, at the age of twenty-two when he bailed out of a Bf 109G-2 after the reduction gear to the propellor fractured and caused a fire. His cockpit filled with smoke and it is believed this caused him to bail out badly and strike the tailplane of his aircraft.

The variant models of the Bf 109 which this phenomenal pilot flew during his relatively brief combat career presented no adjustment problems in his unbelievable flair for deadly accurate deflection firing. Records reveal his armorers maintained a rather accurate accounting of the rounds he expended in battle and they averaged a conservative fifteen per kill.

Enemy or not, Marseille was a fighter pilot's pilot. And Willi Messerschmitt's formidable Bf 109, when in the hands of men such as Marseille, was most certainly the deadly gun platform it had been designed to be. The 109 in all its variants enjoyed the longest and greatest production run of any aircraft; more than 33,000 of the type were built. All of Germany's top aces flew the 109 and this aircraft produced more aces than any other. The 109 is also credited with more aircraft destroyed than any other fighter type produced.

North American P-51 *Mustang*

On patrol high over Germany, near Dessau, Major Nevin Cranfill led his 368th Fighter Squadron of the 359th Fighter Group into a running air battle with a flight of Messerschmitt Me262 twin-jet fighters. During the encounter he scored his fifth aerial victory and became an Ace. The 368th first jumped a group of three of the speedy German planes that were passing above their squadron of P-51s. Just as the action commenced, Major Cranfill noticed a box formation of B-17s below him and a flight of ten more 262s beginning their dive toward the big bombers. Major Cranfill immediately gave chase to this greater threat and locked onto a 262 that had made its pass through the formation and was in pursuit of a P-51. After severely damaging a wing on the jet, Major Cranfill lined up on another 262 and commenced firing from 800 yards as he closed in for the kill. The Messerschmitt began to come apart as it banked to the left. Cranfill watched as the aircraft slipped into a diving turn, then plunged to the ground and exploded.

Mustang victories over the much faster jet became as common as those over the other German fighter planes after *Mustang* pilots gained a little air-to-air experience on the jet's performance characteristics. By the war's end practically all of the Me262's produced had fallen victim to the guns of *Mustangs.*

The *Mustang* will undoubtedly be regarded as a legend in the annals of aircraft, a status achieved by only one or two other aircraft that fought during World War II.

But it was not the *Mustang's* performance against the Messerschmitt Me262 that earned it the honor of being known as the best all-round fighter produced during World War II. Long before the twin-jet made its presence known among our bomber formations, *Mustang* pilots the world over had racked up a formidable tally of aircraft kills, plus highly destructive and significant attacks on ground targets and shipping. In the end, it was P-51 pilots who, after finding dwindling aerial targets, actually stalked airfields in Germany—hunting up something to hit.

Conceived by the British and designed and built by the fledgling North American Aviation Company, the U.S. Army Air Corps took the aircraft that had been ordered by Britain. The legendary P-51 rolled off the line in only 117 days after "go-ahead," three whole days ahead of schedule. With exceedingly high performance characteristics, easy to fly, and relatively forgiving, the P-51 quickly became a favorite among fighter pilots.

From entry into action during World War II until phase out from action in Korea, the *Mustang* gave a majestic and memorable account of herself.

Yakovlev *Yak*

The *Yak* in all its variants enjoyed a long and lusty fighting career. Introduced in May of 1940 this "workhorse" of a fighter saw service until 1953. By most estimates the *Yak* ran a close second to the Messerschmitt Bf 109 in numbers produced.

Below 16,000 feet the scrappy fighter was a terror to the German Luftwaffe. At lower altitudes the *Yak's* superior speed and maneuverability undoubtedly was a great assist to her many pilots who racked up enviable scores of downed 109s. And Russian pilots were fond of getting right on the deck and busting tanks with the *Yak's* nose mounted cannon.

Russians were not the sole users of the *Yak* fighter. The Free French, Polish, Yugoslav and North Korean air forces also employed the aircraft. It was the latter who dashed across the 38th parallel with half a dozen *Yaks* and destroyed a C-54 near Seoul, Korea. A couple of days later *Twin-Mustang* pilots tagged three of the *Yaks* ending their presence in Korea. This encounter in 1950 was not the first clash between *Yak* and American planes; it was, however, the last, and in both instances the little fighter that had punished the German so well came out a poor second against the American.

The first engagement occurred in early November 1944, over Yugoslavia. Russian ground forces had the German in retreat. The 15th Air Force was requested to provide close air support. Colonel C. T. "Curley" Edwinson's [now General (Ret)] 82nd Fighter Group operating from Foggia, Italy, caught the mission. The husky P-38's performance was so good that the Russians asked for a repeat support mission to be flown by the same group on the following day. Again, Edwinson led his three squadrons of P-38s across the Adriatic and down into the valleys of mountainous Yugoslavia. Unknown to Edwinson a crisis was in the making. The Russians had failed to advise Foggia that during the interval between the previous day's support mission and now, Russian ground forces had advanced the battle line by 100 kilometers. Edwinson led the P-38s into the strafing attack that ripped first into the Germans then immediately into the Russians. The resulting devastation was both massive and effective. Caught in the strafing was a Russian staff car. Its occupant, a three-star General, was killed, a victim of lack of communications and a close similarity between German and Russian uniforms and vehicle color schemes. And with the P-38's speed these differences went unnoticed. A flight of *Yaks* were in the vicinity and the call went out for them to attack the P-38s still busy making strafing runs. Caught totally by surprise, Edwinson saw two of his aircraft being shot down. Instantly he signalled the squadron to disengage from the ground attack and fight their way out of the valley. During the brief air battle that ensued Edwinson's P-38 pilots knocked down four of the *Yaks* and sent the remainder scurrying away into the haze. One of the four *Yaks* that really got it was the unlucky fellow whose course took him directly over the guns of the P-38 piloted by Bill Blurock who was in a stall condition and but a few yards under the Russian. A touch of the button and the *Yak* was literally ripped to shreds. It is this moment of the action depicted in this painting.

This incident over Yugoslavia gave the United States a 4-to-2 edge in the only known aerial combat between the two powers (the 1950 engagement involved North Korean pilots). When advised that the situation was one of those unfortunate happenings that bad communications sometimes foster, and after all it was the Russians who attacked the P-38s, the Russians promptly shot those involved on their end and demanded the same be done to Edwinson, the leader of the P-38s. "Curly" Edwinson was quietly and hastily re-assigned to a base out of Europe.

A bad day for these particular *Yaks* should not be interpreted as a sign of inferior equipment . . . not so. Its overall combat record is excellent. Edwinson had superior fighters and some of the finest pilots around. It was no contest even though the Russian pilots entered the action with the advantage of surprise and at a fighting altitude of their choosing.

Republic P-47 *Thunderbolt*

Probably the last item of the design criteria for this brute of an airplane was that relating to ground support missions.

Big, barrel-shaped (the P-47 was the biggest, heaviest fighter produced by America during the war), this rugged airplane was so well built and armored that all of her leading aces survived the war. The mighty "Jug," as the P-47 came to be known, was notorious for getting you back. Not only could the "Jugs" dish it out in a real slugfest, but they could also take a terrible beating and keep on flying.

At one time or another during its World War II service life, the P-47 *Thunderbolt* was assigned to slightly less than 60 fighter groups, an impressive dispersal record. Yet, four Air Forces (4th, 6th, 11th and 13th) flew no P-47s in any of their fighter squadrons. It was the 9th Air Force that was to be the biggest user of the *Thunderbolt,* sixteen fighter groups.

While the heavy bombers of the 8th Air Force busied themselves with the high altitude precision bombing of Hitler's industrial cities, it was the medium and light bombers of the 9th Air Force which fell heir to the destruction of his bridges, barges, marshalling yards and long lines of ground combat forces. Both Air Forces used P-38s, P-47s and P-51s primarily to protect their bombers as they went about their bombing missions. And the fighter pilots did a remarkable job, as the records show.

Because the 9th operated down closer to their targets and because the fighters of both the 8th and 9th, along with their Allied partners, had done a rather thorough job in cleansing the air of the once-dreaded Luftwaffe, it followed that the day would come when the fighters would turn fighter-bombers and go down on the deck to help root out the dwindling German forces there.

Late September, the 29th of 1944, provided most of German activity for the 9th Air Force's P-47s. Hitler began moving the best that was left of his once-proud armies into the Ardennes Forest, slowly, inconspicuously moving them through the byways and waterways but always towards the Ardennes where the queen of all land battles would be fought —the Battle of the Bulge.

Screaming down out of the sky to shoot up everything that moves and some that does not, are 9th Air Force, 373rd Fighter Group P-47s. They get what they can at this river site before terrible weather shuts them off from the action at the Bulge.

Many countries had this truly fine fighter-bomber in their inventories. Among them were numerous Central and South American nations including Peru, Chile, Mexico, Nicaragua, Bolivia and Ecuador to name a few. And, of course, the big users such as the Royal Air Force, China, and Russia all had the mighty "Jug."

Lockheed P-38 *Lightning*

Admiral Isoroku Yamamoto, Commander in Chief of the Japanese Combined Fleet, was a punctual man. It was this discipline in his character that cost him his life the morning of 18 April 1943, Palm Sunday.

Yamamoto sat in the cockpit of one of two *Betty* bombers flying a southeasterly course down the coastline of Bougainville. Above and in front of him, flying into the sunrise, was an escort of six *Zero* fighters. The group had lifted off from Rabaul at precisely 0600 hours and was now only minutes away from its destination, Ballale Island, where the Admiral and his staff would begin an inspection and morale boosting tour of the forward defenses. In strict compliance with the Admiral's schedule the flight would arrive at 0800 hours.

Four days earlier intelligence personnel at Pearl Harbor had intercepted and broken the coded Japanese message that revealed the Admiral's planned movement and the limited aerial strength. After hasty deliberations within the political and military complex, the decision was made to intercept and destroy the Admiral.

The mission was assigned to the 339th and 12th Fighter Squadrons flying P-38 *Lightnings* from Fighter Strip Number Two on Guadalcanal. Extra large auxiliary fuel tanks were moved in during the night of the 17th and installed on eighteen *Lightnings*. It was a long 650 miles from Guadalcanal to Bougainville.

As dawn broke on the 18th, the heavily laden *Lightnings* used all of Strip Two in getting airborne. In the process one P-38 blew a tire and skidded off the runway. Then, as the remaining P-38s formed up, another one suffered fuel system problems and was waved off. Sixteen *Lightnings* now continued toward Bougainville to meet a destiny that would in its own way alter the course of the war.

Twelve of the sixteen craft were to provide defensive cover for the other four which had been designated as the killer group. Their sole objective was to shoot down the two Japanese bombers—regardless!

Right on Yamamoto's schedule the two groups met. Jettisoning their tanks, Tom Lanphier and Rex Barber throttled in for the kill as Besby Holmes fought frantically to release his tank which stubbornly refused to let go. Raymond Hine, seeing Holmes' vulnerability, aborted his attack on the *Bettys* to fly wing position and protect Holmes until the tank problem was solved.

From beginning to end the action consumed a bare thirty seconds. In that brief course of time both of the Japanese *Betty* bombers were downed along with three of the escorting *Zero* fighters, all from the guns of the four killer group *Lightnings*.

Shown here is a glimpse of that fight with Tom Lanphier's P-38, "Phoebe," in the foreground. Behind him is Yamamoto's *Betty* breaking up prior to crashing in heavy jungle growth. Below is the other *Betty*, which after having been virtually chopped to pieces, crashed heavily into the water just offshore the Japanese base at Buin. Miraculously, three survived the crash of the second *Betty*—Vice Admiral Ugaki, Rear Admiral Kitamura, and the pilot, Warrant Officer Hayashi.

One American, Ray Hine, failed to return.

It is significant that after this brief but devastating engagement the Japanese never again achieved a major victory, only disasters that crept relentlessly toward the home islands. Yamamoto, the great naval strategist, was no more.

Bell P-39 and P-400 *Airacobra*

When the Japanese struck Pearl Harbor, the United States' first-line of fighter defense was composed of Navy and Marine piloted F4F *Wildcats* and the Army Air Force piloted P-40 *Warhawks* and P-39 *Airacobras*. To these rugged but obsolete aircraft fell the tough task of buying us the time needed to gear up for war.

Designed to fight a completely different generation of enemy aircraft, two of the three fighter types would emerge from the early onslaughts with good reputations; the third, the *Airacobra*, would find the going pretty tough. Designed to fight at altitudes under 15,000 feet, the sleek long-nosed P-39 was soon weighted down with so much fire-power capability that by the time a fighting altitude could be reached the battle was invariably over, and in the Pacific the *Zero* pilots simply climbed beyond the P-39's ceiling then pounced. As radical and innovative as it was, the P-39 simply could not hold its own against the Japanese *Zeros* and German *109s*.

The British were probably the most discouraged recipients of the P-39 (P-400s as the export version) while the Russians were unquestionably the happiest. Leaving the heroics of dogfighting to other aircraft types, the Soviets put the P-39 down on the deck where it belonged and where it could best hurt the German.

As MacArthur began the frustrating process of making good his promise to "return" from his Australian headquarters, crated P-400s destined for the British were rerouted to Tontouta, New Caledonia, for use by Americans who had no planes, but were now at war. As the men labored in the mud and rain to uncrate and assemble the forty-five P-400s, they discovered nobody had ever before met one. Obstacle after obstacle was overcome in the rush to assemble the craft, a feat accomplished without benefit of any technical manuals in only six days. The 67th Fighter Squadron could now say it was ready to fight.

In time the 67th moved nearer to the action and began operation from Guadalcanal. It was here they found by trial and error the strengths and weaknesses of their mixture of P-400s and P-39s. Reluctantly, the 67th let the derring-do of fighter scrambles go to the *Wildcats* and *Corsairs*. As we see here, they found their power to hurt the enemy was ripping into him down on the ground, appearing suddenly from nowhere, tearing to shreds anything standing in its way. The P-39s and 400s of the 67th were easily recognized by the sharkmouths they took from the departing P-40s. P-400s were left in the British camouflage scheme that was on them when they were uncrated in New Caledonia. Only the British markings were painted over.

As a part of the 374th Fighter Group the 67th transitioned into P-38 *Lightnings* in August of 1944. They were the last Fighter Group to fly *Airacobras* in the Pacific.

Grumman F4F *Wildcat*

On the 3rd day in May of 1942, three days prior to the fall of Corregidor, Admiral Yamamoto sent Vice Admiral Hara out of Rabaul and into the Solomons and Coral Sea in force. The three-fold objective was to take Tulagi on Florida Island in the Solomons and establish a seaplane base, then take Port Moresby situated in the southeastern part of New Guinea, and, finally, if possible, to draw out the remaining American naval power and destroy it once and for all.

On the first day of May, Task Force 17 under Rear Admiral Fletcher had entered the Coral Sea from the southeast.

American cryptanalysts had long since broken the Japanese code and therefore clearly understood the enemy's intentions as they ventured into the Coral Sea and Solomons area.

Now would commence the strangest of all sea battles, The Battle of the Coral Sea. For the first time in maritime battle history, not a single shot would be exchanged between surface vessels. It would be a decisive naval engagement fought entirely by the opposing forces' aerial strengths.

On 3 May Tulagi fell to the Japanese, precisely as scheduled by Yamamoto. On May 4, *Yorktown* planes made several retaliatory air strikes on Tulagi that proved to be of little consequence. The 5th and 6th were quiet as both sides swept the Coral Sea in search of another, especially for the carrier forces which were known to exist. Then, on the 7th, contact was made with the Japanese invasion force headed for Port Moresby. Immediately, *Yorktown Wildcats,* dive bombers, and torpedo planes raced to the attack; and for the first time the word flashed back "scratch one flattop." Within ten short minutes the carrier *Shoho* slipped beneath the waves of the Coral Sea. With the sinking of *Shoho,* any further advance toward Port Moresby was aborted by the enemy.

But it was the fight on the 8th of May which would show the muscle that foretold what the outcome of the future naval battles would be. On this day the giants would clash.

It was an evenly matched battle with the slight weather advantage for the Japanese. Both sides had equal numbers of aircraft, carriers, and supporting warships of the line. The two Japanese carriers, *Zuikaku* and *Shokaku,* had participated in the attack on Pearl Harbor. Opposing this pair of leviathans were Task Force 17's heavy carriers *Yorktown* and *Lexington.* Aircraft from both forces found each other's carriers about the same time, and without hesitation attacked. *Zuikaku* escaped into heavy weather, but *Shokaku* was less fortunate; *Yorktown* dive bombers placed two five-hundred pound bombs on her deck, warping it, and a later strike added a third. Beyond the horizon both the *Lexington* and *Yorktown* were sustaining hits; the *Lexington's* damage was of such a nature that salvage was impossible, and she was subsequently sunk by Task Force 17 destroyers.

The seizure of Tulagi proved to be no major threat to our later movements around New Guinea, the Bismarck Barrier or the Solomons. A bashful Japanese Admiral had turned his invasion force away from Port Moresby, never again to threaten the place with such a glorious opportunity; one enemy carrier, small as it was, had been sunk, and the two large carriers were to remain non-operational in Japanese home waters and miss the momentous Battle of Midway.

Rising up out of the flak from the damaged but fighting *Shokaku* are two F4Fs from VF-2 off the *Yorktown.* It was this fighter type along with the world-famous sharkmouthed P-40 *Warhawks,* P-39 *Airacobras,* and British *Spitfires* that held the line fighting off the best the enemy had until the next generation of fighters could come off the production lines to enter the battle arenas.

In the decisive Battle of the Coral Sea, twenty-two F4F *Wildcats* of VF-2 operated from the *Yorktown* and twenty from VF-42 rose from the deck of the *Lexington.* The Battle of the Coral Sea proved the Japanese could be stopped. And air power did it!

Grumman F6F *Hellcat*

Occasionally, a truly superb combat aircraft fails to receive the high degree of recognition that it rightly deserves. Such is the case of the mighty Grumman F6F *Hellcat,* successor to the F4F *Wildcat* and bearing a close family resemblance to it. It was a dedicated competitor of the fine F4U *Corsair.* No question about it, the powerful F4U was "whistling death" to the Japanese. The record, however, reveals it was the deadly F6F that clobbered the aerial might of the Japanese.

Fate moves in strange ways. And so it happened that Yamamoto's northern force that struck the Aleutians during the Battle of Midway allowed a *Zero* to crash land intact. The craft was quickly shipped stateside where it was reassembled. It was subjected to severe test flights to reveal its weak and strong flight characteristics. The F6F *Hellcat* was the fighting machine designed specifically to counter the *Zero's* strong points and take advantage of its weaknesses. It was superior to the best the Japanese could put into the air. And from its initial debut in the Marcus Islands this fighting plane proved its superiority over Japanese aircraft.

Statistically, *Hellcats* are credited with the destruction of more than 5,000 enemy aircraft while the *Corsair's* score is slightly over 2,000. This is in the face of the fact that more *Corsairs* were produced than *Hellcats* and, more importantly, entered the Pacific action at least six months prior to the *Hellcat.* Both aircraft racked up fantastic records.

From the onset *Hellcats* had an insatiable appetite for enemy aircraft. Not only did they turn in noteworthy performances in the Gilberts and Carolines, but they baited, then destroyed everything flyable in and around Truk, the impregnable Japanese Pearl Harbor of the Pacific.

Came the dawn of 19 June 1944 and the stage was set for what was to become known as the "Marianas Turkey Shoot." Separated by some 400 nautical miles of Philippine Sea were Vice-Admiral Ozawa's forces and Admiral Marc Mitscher's Task Force 58, tremendous armadas of naval surface warships and airpower. When the opposing forces found one another, the Turkey Shoot began. By sunset *Hellcat* pilots had bagged nearly 400 enemy aircraft with a loss of only eighteen *Hellcats*, a lopsided victory if ever one existed and a great testimonial to the F6Fs' fighting abilities and the tenacity of its pilots.

Depicted here is the F6F-5, "MINSI II," flown by the Air Group Commander of VF-15 (Fabled Fifteen) off the carrier *Essex*. It is Commander David McCampbell enroute to racking up nine enemy aircraft over the Philippines on 24 October 1944. Part of Commander McCampbell's Congressional Medal of Honor citation for this exploit reads ". . . During a major Fleet engagement with the enemy on October 24, Commander McCampbell, assisted by but one plane, intercepted and daringly attacked a formation of sixty hostile land-based craft approaching our forces . . . shot down nine Japanese planes and completely disorganizing the enemy group, forced the remainder to abandon the attack before a single aircraft could reach the Fleet . . ."

And he kept score on his instrument panel with a pencil.

Mitsubishi A6M5 *Zero*

A Pacific dawn broke silently over Iwo Jima on 24 June 1944. The low overcast of a leaden sky was a welcome treat to the Japanese garrison manning the defenses of the tiny island. Their brief respite from the harassing American bombers and warships was to be short-lived, however. Almost immediately after 5 a.m. the familiar sounds of air raid warnings echoed over the island—a signal for the 80 defending *Zero* fighters to scramble and meet the attackers. Amid the klaxoning sirens they rose to swiftly form up tightly just beneath the overcast. The sky was void of Americans. On signal about half of the *Zeros*, led by veteran pilot Kinuste Moto, nosed up into the overcast. Breaking out into clear skies he found nothing. Then, suddenly bursting through the clouds, came a loose formation of Navy *Hellcats*. Immediately Moto attacked and in seconds had personally blasted four from the sky. Stunned by the surprise, the Americans throttled back into the cloud cover and on popping out ran headlong into the lower echelon of *Zeros* searching for

them. Among these Japanese pilots was Saburo Sakai, an Ace in his own right. Sakai, during an earlier engagement, had lost his left eye which restricted his vision so vital in the fast-breaking situations of aerial warfare, especially dogfighting with superior machines such as the F6F. Nonetheless, Sakai and his group, assisted by the *Zeros* that had plunged back down from above the cloud cover, ripped into the American force and began separating it into easy "kill" segments. The *Hellcat* pilots, however, were fast to analyze the tactic and jumped to the offensive.

Saburo Sakai, strained by a long layoff from combat and hampered by his limited vision, broke off from the ensuing fast-paced action for a short breather before resuming the fight. Climbing back into the waning battle, Sakai slipped into position with what he surmised to be fifteen friendly *Zeros*. Closing on the group, he suddenly spotted the easily-recognized stars and bars on the *Hellcats*. Instinctively, he rolled away hoping they had not noticed his approach. Looking back, Sakai saw the *Hellcats* banking to the attack. Suddenly the impossible took place. Saburo found himself circling inside a wide circle of *Hellcats* whose pilots took turns breaking from the ring to attack him. Trying every evasive maneuver he had ever learned, none succeeded in shaking the *Hellcat* pilots, and his rolling dives that commenced at 1,600 feet suddenly put him on the deck. Ahead of him stood towering and dangerous cumulus cloud. Behind him were the *Hellcats*. Flying just above the waves (three times a wingtip cut a wave), Saburo raced headlong into the cumulus cloud. Tossed in every direction, the light *Zero* was suddenly thrust upside down into clear skies on the far side of the turmoil. Righting his plane, Saburo spotted the *Hellcats* regrouped on a course taking them away from Iwo.

A portion of the above-described action is captured on canvas, as Saburo makes his life-or-death high-speed escape run from hotly pursuing *Hellcat* pilots.

In this aerial engagement forty of Iwo Jima's eighty defending *Zeros* were shot down while the Americans lost ten *Hellcats*, a respectable 4:1 ratio. During the second engagement twenty additional *Zeros* fell to the guns of the powerful *Hellcats*. A third battle reduced the defending force by eleven more *Zeros*, leaving a token nine battle-weary fighters to defend Iwo Jima.

Chance-Vought F4U *Corsair*

During their desperate and futile bid to reverse the trend of action during the Battle of Leyte Gulf, the Japanese introduced to the Navy and the world a new word—*Kamikaze!*

Conceived out of necessity, the tactic was brutally simple. Faced with rapidly dwindling airpower (and virtually no experienced pilots), the Japanese could no longer mount the large scale aerial attacks that had once terrorized all areas of the Pacific and Southeast Asia. Instead of the conventional mass attack, it was now the lone fanatical *Kamikaze* pilot who flew his aircraft headlong through defending fighters and then heavy walls of flak to a suicidal death as he slammed into the largest warship he could see, usually a carrier.

Shortly after the introduction of the *Kamikaze* in late October, 1944, and before the invasion of Lingayen Gulf at Luzon in early January, 1945, the fast carrier *Essex* took on board the first carrier based *Corsairs* of the Marine Corps, VMF-124, and VMF-123.

Patrolling in the Sulu and China Seas, the *Essex's Corsairs* provided air support as the Americans took Mindoro Island, a stepping-stone enroute to Luzon, then began interrupting the enemy's attempts to beef-up their Lingayen Gulf defenses. In these seek-and-destroy missions what little Japanese air strength remained within striking distance of our invasion route and the Luzon objective was methodically searched out by the *Corsairs*. Not only were *Corsairs* seen in the Philippines but also over Formosa and Saigon. And so it was that during this time the *Kamikazes* of the Special Attack Corps found not only the *Essex* but also the *Cabot, Hancock* and then, "Evil I," as the forever damaged *Intrepid* came to be known.

The scene is somewhere over the Philippines with *Corsairs* of VMF-124 breaking into a swarm of the suicide bent *Kamikazes*.

It would be an affront to this marvelous "bentwing" airplane and her many pilots and maintenance crews not to mention just how great the design really was. From a maiden flight in May of 1940, this powerful machine remained in "first-line" status until the 1960s, the longest service life of all fighters. As a propeller-driven craft the mighty *Corsair* outlived all other American-built fighters, land or carrier based.

Grumman F8F *Bearcat*

Bearcats brought to the table the best features of their older sisters, the F4F *Wildcats* and F6F *Hellcats,* plus a new power all their own.

Roughly the same size as the rugged *Wildcat* and possessing the protective armor of the deadly *Hellcat,* and with a magnificent power-plant, the F8F promised to be a winged hell in battle with the Japanese.

During February of 1945, *Bearcats* aboard the U.S.S. *Charger* pulled off carrier tests with excellent results.

In May the first *Bearcats* entered into service as VF-19. June and July saw the VF-19 brought up to strength (nearly fifty aircraft) while the squadron's pilots and crews underwent an exhaustive training program. Then, on 2 August the U.S.S. *Langley,* with VF-19 aboard, cast off all lines and headed for the far reaches of the western Pacific where the battle was raging ever closer to Japanese soil. Two destructive blasts at Nagasaki and Hiroshima gave the enemy a quick face-saving way out of the war. The *Langley* and her *Bearcats* of VF-19 never made it into action.

V-J Day did not, however, bring a halt to the F8F program as it did to so many others in progress. Production and pilot training of the *Bearcat* continued. In mid-1946 it was determined that F8Fs and F4Us would have to hold the line until the new jets could be brought into service. And it was about this time that the U.S. Navy selected the F8F to be the plane for their first "Blue Angels."

Came Korea in 1950 and the nod went to the F4U instead of the *Bearcat*. F4Us were known to be workhorses in support of the foot soldier. *Bearcats* were designed to be fighters. It appeared that the *Bearcat* was destined to go the route of the monstrous B-36 and be denied the taste of battle.

Finally, on her third chance, the *Bearcat* got the call. She would fight in French colors and markings and right down on the deck where she had been denied the opportunity in Korea. The French in Indo-China, Vietnam in particular, used the F8F very effectively as a ground support aircraft even though the famous fight-to-the-finish stand at Dien Bien Phu was lost and brought an end to the French presence in that part of the world.

South Vietnam received the remaining French *Bearcats* when that country was separated from the North. Later, Thailand received F8Fs as a result of a U.S. Military Air Advisory Group program to strengthen that country's air power.

But all of that was long after this scene which catches a pair of *Langley* F8F *Bearcats* patrolling a naked Hawaiian sky in August of 1945—shortly before World War II closed its books.

Nakajima Ki-43 *Oscar* (*Kamikaze*)

The Japanese word *Kamikaze* means Divine Wind. Its significance in Japanese history dates back to 1280 when the great Mongolian warlord, Kublai Khan, sent a huge invasion fleet to take the Japanese islands. In a situation similar to that of the Spanish Armada in the English Channel, as the mighty Chinese naval force lay off the Japanese coastline a killer typhoon moved in and wrecked the fleet. To the Japanese this was indeed a *Kamikaze* sent by the Sun Goddess to save them from certain annihilation. The Divine Wind destroyed the Mongol naval threat and the Japanese people survived.

Centuries later came the great Pacific war and the mighty United States Navy. The two great sea powers tugged at one another until the climax came at the Battle of Leyte Gulf. The Japanese Imperial Navy moved into the Philippines in force to have it out with the Americans, once and for all. The clash developed into a lop-sided action with the U.S. Navy emerging as the clear-cut victor.

For some time prior to the Battle it had been recognized by some Japanese naval officers that their air fighting effectiveness was all but gone and in order to handle the powerful American fleets a new and innovative tactic was needed. The answer lay in the suicide mission, where a 130-pound man in a single engine plane carrying a 500-pound bomb might be traded for a giant aircraft carrier or cruiser. Thus, in 1944 the first *Kamikaze* Special Attack Corps was formed and its pilots entered action for the first time during the last days of the Battle of Leyte Gulf. And these *Kamikaze* attacks were successful. Emotions ran high among some Japanese pilots to become a *Kamikaze*. They believed they would be assured godhood at the Yakasuni Shrine if they gave their life for the emperor and country.

The successes at Leyte sent even greater numbers of the pilots into the action during the January 1945 invasion of Lingayen Gulf, and the *Kamikaze* attacks reached their peak during Operation "Iceberg," the invasion of Okinawa in April of 1945.

The once mighty Japanese Imperial Navy was now gone from the scene and with it practically all of her good first-line pilots. Japan faced an enemy armada of more than 1,500 ships that were knocking on the very doors of their homeland, a desperate situation not unlike what they had faced with the terrible Khan and his fleet. It rested with the *Kamikaze* pilots to destroy the American fleet. They were the only meaningful Japanese defense that remained.

The effectiveness of the *Kamikaze* was there; at least 40 American ships were sunk at Okinawa and more than 350 damaged. Had the tactic of *Kamikaze* missions been started earlier in the war, and been better organized, it would have delayed the United States timetable of events and caused the overall naval strategy to be reconsidered.

No ship was safe from the *Kamikaze*, but the aircraft carriers were the primary targets of the young pilots. The big U.S.S. *Franklin* fell victim during the Battle of Okinawa. Our painting shows another dedicated *Kamikaze* pilot driving his *Oscar* toward the burning, listing, but still fighting, warship. Will he make it to the fighting *Franklin*, will he be able to crash through the blistering firepower thrown up at him? When the action was over the *Franklin* counted some 700 dead and more than 250 wounded, and she was towed out of the battle by the cruiser *Pittsburgh*.

HEAVY BOMBERS

Boeing B-17 *Flying Fortress*

This tough "Queen's" battleground was anywhere the enemy elected to show himself. *Fortresses* will, however, always be remembered for their magnificent performance in disrupting the mighty Axis war machine. Massive daylight precision bombing attacks from high altitudes left Hitler's roofless European bastion in shambles. But that was later on. Before Hitler's legions would feel the might of the B-17 *Flying Fortresses* of General Ira Eaker's 8th Air Force, the rampaging Japanese would first have to face this aircraft in action.

On the third day after Pearl Harbor and after being all but decimated at Clark Field outside Manila, B-17 remnants of the 19th Bombardment Group made the first American aerial assault strike of the war when they bombed Japanese shipping near Vigan, Luzon.

After being whisked away from Corregidor and sped to Mindanao in PT boats, General MacArthur and his entourage continued the reluctant journey to Australia by B-17 *Flying Fortresses.*

It was from Australia that MacArthur would gather together what men and equipment he had, mold them into a fighting monster, and begin the long trek to Tokyo in consort with Admiral Nimitz. A part of MacArthur's seemingly impossible task was the resurrection of a highly disorganized and demoralized air force.

Tales of unimaginable heroics and acts of great courage by B-17 airmen of "Blondy" Saunder's 11th Bombardment Group (H) during the time spent in containing the resolute Yamamoto, in his design for Australia and destruction of the U.S. Navy, are far too lengthy for an accounting here.

Selected to exemplify the courage and dedication of all *Fortressmen* and to refresh the memory of all as to the structural strength of this majestic airplane are a duo of 13th Air Force, 98th Squadron B-17Es, "Galloping Gus" and "Typhoon McGoon II," hooking it home after tearing up a piece of New Guinea real estate temporarily held by Japanese garrisons. Rising up to do battle with them are slow moving *Rufes,*

easy marks for the *Fortresses* gunners. At the controls of "Typhoon McGoon" is Captain Walter Yates Lucas, Squadron Commander of the 98th.

March, 1943, was the beginning of the end of the *Fortress* presence in the Pacific. After fighting the decisive Battle of the Bismarck Sea, General Kenney (5th Air Force) went to Washington for the express purpose of getting more planes, especially B-17s. General Arnold denied his request for the B-17s and Kenney went straight to the president with the request. Again, he was told all B-17s were earmarked for Eaker's Eighth Air Force in Europe. Heavies in the Pacific were to be *Liberators* and the B-17's big sister, the gigantic B-29.

Boeing B-29 *Superfortress*

The time is Wednesday the 23rd of May, 1945. Curtis LeMay's 20th Air Force B-29 *Superfortresses* are attacking the Japanese homeland again . . . this time the target is Tokyo's urban area. Four times previously Tokyo had been visited by the giant bombers that rained down fire and destruction. Tonight's mission (#181) and one more on the 25th of May (#183) are to be the final punches that will destroy most of Tokyo, reducing the area to a barren sea of ash.

Laboring northward from their four Marianas bases, 558 *Superfortresses* (the largest number of B-29s over a single target in WWII) from the 58th, 73rd, 313th and 314th Bomb Wings trudge for Japan; this long irregular chain of aluminum undulates gently under the glow from a three-quarter moon hanging in the southeastern sky. All eyes turn forward as the word is passed: *"Fuji . . . dead ahead!"* The familiar landmark looms in the distance, its majestic snow covered peak reflecting a dull red cast on the side facing Tokyo over fifty miles away, telling everyone that the pathfinders and lead bombers have made Tokyo a raging inferno.

Fuji is the signal to the crews to become even more alert. Ahead, only minutes away, they will make landfall, fix their position, then turn and take a new course for the run to target.

At least eighty-three attacks by enemy aircraft are made on the *Superforts* this night; the anti-aircraft fire is blistering hot, especially from Kawasaki and Yokohama to Tokyo. A seemingly endless corridor of searchlights spear the clouds, diffusing themselves to meld with the patterns cast by the fires of Tokyo. *Kamikazes* plow head-on into the armada of B-29s while *Betty* bombers launch rocket powered *Baka Bombs* that blaze in from above the *Superfortresses;* most will take their one-way ride on a long, arching plunge to a collision on the ground or into the bay; a few, however, will connect and the sky is rent with pyrotechnics.

When it's all over two days later, Tokyo is stricken from the list as a target. More than fifty-six square miles of the city has been gutted by fire. All that remains upright are a few lucky smoke stacks and street light standards. The heavy stone bridges and paved streets are there, but all else in the target area is a barren waste.

Tokyo was one of sixty-five principal cities destroyed by the 20th Air Force. Five other great industrial areas were also reduced to rubble: Nagoya, Kobe, Osaka, Kawasaki, and Yokohama. The effect of the B-29 on Japan was overwhelming and certainly shortened the war and saved thousands, if not millions, of lives on both sides. The B-29 displaced an estimated 21,000,000 persons, and killed and wounded more than 800,000, which were more casualties than the combined Japanese armed forces suffered during the three and a half years of war with the United States.

Mission 181 was selected to portray the *Superfortresses.* In the foreground is the B-29 "Eddie Allen," a veteran of ten trips over the "Hump" and more than twenty missions against the enemy; Mission 181 was its seventh from the Marianas. This 58th Wing aircraft, a gift from the Boeing people, was named in memory of the Boeing test pilot who did much to develop and approve the B-17 and B-29, and who perished while trying to save the second B-29 prototype during a test flight in December 1942.

"Eddie Allen" has made its bomb run on Tokyo. Coming off target we see Captain Eino Jenstrom and his crew bank the mortally wounded aircraft away from the core of the huge Tokyo bonfire, away from the blinding lights and blastings of anti-aircraft bursts. Fire rages in the outer wing around the fuel cells where a dud round has slammed into the wing next to number one engine. This dud and the fire it caused will weaken the main spar structure and make the great airplane unsafe to fly again. This, then, is the final moments of the "Eddie Allen."

The "Eddie Allen" left an enviable combat record, as did virtually all *Superfortresses.* During eight months of combat, the big bomber hit targets in seven countries: Burma, Malaya, China, Manchuria, Formosa, Thailand, and Japan.

Consolidated B-24 *Liberator*

At dawn on Sunday, the first day in August, 1943, 1,725 Americans in 178 bomb-laden B-24s rose from their airdromes in North Africa and began the assault that would be forever remembered by a single dynamic word, *PLOESTI!*

The stakes for Ploesti were high. The effort expended toward the destruction of the place was so great that this "Queen" of all *Liberator* actions produced five Medals of Honor.

Five Bomb Groups made the attack. One of these Groups was the 98th Pyramidiers which was led by the aggressive and audacious John R. "Killer" Kane. The Pyramidiers, with 47 severely weathered B-24's, drew the largest and most important target area in the Ploesti oil refinery complex—Astro Romana, code named "White Four."

For the assault on "White Four" Kane divided the Pyramidiers into five echelons, four composed of ten aircraft each and the fifth flying short with seven. Kane was centered in the first or leading flight and flew the lead position in John Young's *Liberator,* "Hail Columbia." Two planes away to his right was Lieutenant Royden LeBrecht piloting "The Squaw," the principal aircraft in the painting.

It was a long flight to Ploesti—better than six hours; plenty of time to let the mind recap all the courses and speeds and altitudes to target, more than enough time to mentally review that specific target site one more time. Halfway to target, Arens flying right wing to Kane, signals fuel troubles and aborts from the mission. LeBrecht moves "The Squaw" into the empty space beside "Hail Columbia."

Suddenly Ploesti explodes into life beneath the 98th with the slamming of rounds from German 88s being fired at point-blank ranges.

The situation worsens . . . rapidly. Another Group, Liberandos, miss a key turn coming off target and head into the on-rushing Pyramidiers roaring into Ploesti as planned. Walls of fire and mountains

24

of heavy black smoke hide the specific targets so carefully memorized by the 98th's pilots. Kane, bearing down on all but obliterated Astro Romana, is faced with a crisis decision: continue or abort? He signals to continue the run to target. The 98th is met by steady anti-aircraft fire that is more intense and accurate than advised during briefing. The big bombers charge through the enemy's outer defenses then lunge past the on-coming Liberandos. Hurtling into the broiling inferno the lead pilots open up with their new nose-mounted 50 caliber machinegun clusters. A blossom of flame flashes from the bombers' noses and a fusillade of fifty caliber slugs blast a wide swath before them. Kane leads his ragged looking Pyramidiers into the boiling fireballs and thick oily smoke, past dangerous high-risen smoke stacks and by the menacing barrage balloon cables. After forever the bombs are finally released—and they explode prematurely for the low flying bomber crews. After spending an eternity of a few seconds in the hell of Astro Romana the 98th breaks into the clear on the far side of the furnace and begin to collect themselves for the sprint from the target area. The Pyramidiers come off target hurt, badly hurt, but behind them sprawls a blazing, bursting Astro Romana. "Tidal Wave," the code name for the overall assault on Ploesti, has done in the span-time of only a few precious moments what would have taken an Army of foot soldiers months and many dead to accomplish.

A new danger arises: "Hail Columbia's" number four engine feathered since run to target, shudders and slows speed . . . easy prey for a fighter. Kane quickly gathers in a couple more cripples (they are easy to find) and with the undamaged "Squaw" flying top cover to fend off any fighters takes a heading for the closest Allied airfield which is located at Nicosia, Cyprus. It is a long, long way from Ploesti to Nicosia when you are in an ailing B-24, and it would be dusk before the friendly island would come into view. But before Nicosia the little band would pass neutral Turkey, "Hadley's Harem" loses a pair of engines, signals goodbye, banks away and executes a perfect sea-ditching job next to the Turkish coastline below. "Hail Columbia" slows to near stall speed and the battle weary trio of aircraft plod on toward Cyprus Island. As darkness settles around the *Liberators,* snuffing out the misery far behind them, Nicosia appears. LeBrecht piloting "The Squaw" and William Banks at the controls of the other *Liberator* land and make way for the in-coming badly damaged "Hail Columbia." Kane holds the big bomber on line with the runway and in the quickening darkness settles the aircraft down onto what he reckoned would be runway overage only to hit short, shear off all landing struts and crash land. Nobody was injured.

Ploesti took its toll on the rugged Pyramidiers. Most of the surviving B-24s suffered heavy battle damage and/or aircraft system malfunctions while only a few licked minor wounds. Only "The Squaw" came away from Ploesti without sustaining any battle damage or system failures; she was one of only three 98th Bomb Group aircraft that was on the line ready for a mission the day following Ploesti! This saucy little lady, named by Joe Kilgore, her first pilot, was due for a change in regard to battle damage . . . inside her empennage and waist sections remained the scars from more than two hundred bullet and cannon shot holes that Kilgore picked up from a dedicated Messerschmitt pilot during an earlier mission over Naples. In that action her tail turret and waist positions were shot away by the deadly accurate fighter. After 71 missions "The Squaw" had other scars, to be sure; but she had made it into and out of heavily fortified Ploesti without damage. In the end, "The Squaw's" and LeBrecht's crew were selected to return to the United States to participate in a War Bond Tour.

The part of the action selected to be portrayed was "The Squaw" and "Hail Columbia" as they blasted their way into Astro Romana. Here we have heavy bombers in the role of attack bombers and it isn't very often that you hear of heavies rigged with nose mounted machineguns. And the armorer mounted them with sections from bunk beds! This is one of the extremely rare occasions when pilots of heavy bombers actually had guns to fire.

Consolidated *Liberators* saw duty in all Theaters of Operations. The big burly bomber enjoyed the largest production run of any American made aircraft that saw service in World War II.

MEDIUM BOMBERS

Martin B-26 *Marauder*

For a bomber type that was considered too dangerous to fly and subsequently underwent more than one investigation, the B-26 *Marauder* probably tore up more of Germany and the German war machine, wherever it was located, than any other bomber. It is true that this clean-lined craft had a bad reputation. The original short wing configuration quickly got itself a host of ugly nicknames such as *Widow Maker* and *One-a-Day-in-Tampa-Bay* because with the short wings there were accidents during take-off and landings. Requiring a fast take-off and landing speed, all variants of the *Marauder* were rather "hot" machines for most pilots. However, in the hands of an experienced crew, the B-26 proved to be both the menace she was designed to be plus a truly safe airplane. When the final numbers were in, the powerful *Marauder* held the record for the lowest combat loss rate of all bombers, less than one-half of one percent.

While the B-26 gained most of its fame with the 9th Air Force over Europe, it is well to remember that this fine aircraft also performed heroically in other theaters as well. The Battle of Midway was a naval engagement, yet included in the mixture of ancient to new aircraft that were destined to hurl back the Imperial Japanese Fleet were four B-26 *Marauders* that had been hurriedly converted to carry torpedoes. These four *Marauders* along with six *Avengers* of Torpedo 8 led the initial attack on Admiral Nagumo's carrier forces as they drove for Midway.

Exciting accounts of B-26 missions over Europe are bountiful. Selected to memorialize all B-26s, those who built them and those who flew and serviced them, is *Flak Bait*. Here she is pulling up and away after making a run on the familiar "ski-shaped" V-1 rocket launching site below and behind her. *Flak Bait* and her 449th Bombardment Squadron sisters will bank to starboard, cross the channel, and head for home, but not before running the gauntlet of flak on the way out.

This famous bomber flew no less than 202 combat missions and was on the line ready for her 203rd mission when the war ended. In her long battle career *Flak Bait* became a mass of scar tissue—over one thousand holes had been patched. It is no wonder that *Flak Bait* found her way to the National Air Museum in Washington, scars and all.

North American B-25 *Mitchell*

If General MacArthur was to keep his famous promise "I shall return" to the people of the Philippines, then the Bismarck Barrier would first have to be broken.

Arching gently away from the northeast coast of New Guinea sits New Britain Island, the backbone of the Bismarck Barrier separating the Solomon and Bismarck Seas. Strategically situated on that island's northeast tip is Rabaul. With fine, deep-water port facilities, heavily protected by a ring of shore batteries, and no less than five excellent airfields nearby, Rabaul, the fortress, was the key to the Bismarcks.

As the "G.I." and Australians moved up the New Guinea coast from Buna-Gona and the Marines and sailors leap-frogged up the Solomons and Russells, General Kenney advanced his 5th Air Force bases even closer to the Bismarck Sea, New Britain, and Rabaul.

Rabaul, like Truk, her sister citadel in the Carolines, was destined to be by-passed by our invasion forces, but would be completely cleared of all air and naval power in the process.

There were many strikes carried out against Rabaul; the most bloody of all probably being that of 2 November 1943. Intelligence reported Simpson Harbor to be packed with ships. Kenney put nine B-25 Squadrons (all strafers) and six covering Squadrons of P-38 *Lightnings* into the air and headed them out for Rabaul. In addition to the existing massive anti-aircraft defenses guarding Rabaul, the B-25 crews also found heavily armed destroyers. In a strafing run, the pilots would have to bring their planes over Rabaul, bank toward the harbor area, drop down on the deck, and come in fast, blazing away with all ten fifty caliber machine guns firing forward.

This one raid cost Kenney eight bombers, nine fighters and forty-five airmen, but Rabaul, would never quite be the same again. Rabaul itself blazed fiercely as tons of supplies, armament, fuel, and structures went up in smoke. In addition to more than thirty ships sunk or damaged, the enemy also lost nearly one hundred aircraft.

Coming at you in the painting is one of the eighty or so B-25 strafers tearing up Rabaul. Modified in the field by Major Paul I. "Pappy" Gunn, the B-25 "customized" into a strafer placed an enormous concentration of firepower up forward, four fifty calibers in the nose (he removed the bombardier's compartment and the lower turret) and two each mounted on both sides of the fuselage section just beneath the wing. And then there was the top turret gunner who could bring two more into forward firing position during the run to target.

This awesome amount of fire power when used in conjunction with the skip-bombing technique perfected by Kenney's B-25s and his own terror, the parafrags, broke the Bismarck and permitted MacArthur to hasten the pace along the northern New Guinea coast then back to the Philippines, where he did keep the promise.

Dornier Do17

The *"Flying Pencil"*! And what a graceful shoulder-wing bomber was this twin-engine aircraft.

This gorgeous plane made her world debut in 1934. Like the Lockheed *Hudson* and Heinkel He III, the Do17 was originally designed for a civilian purpose. It was originally built to serve as a fast-flying mail carrier that also provided accommodations for six passengers. The first bomber version of the "Flying Pencil" appeared in 1936 and the bomber became operational the following year. In 1938 the Do17 was introduced into the Spanish Civil War fighting and gave an excellent account of itself in that action.

A large number of models were built of the Do17 with the final model being the Do17Z. It entered service in 1939. No other models appeared after the Do17Z.

The Dornier Do17 was a favorite bomber type of pilots and ground crews and the airplane was very reliable in the air. The bomber fell short in its capability to deliver the bigger payloads of its counterparts and as a result was phased out of the fighting in 1942, but not before giving a lusty account of herself.

Dornier Do17s were four-place aircraft and sported three 7.9mm MG 15 machineguns; one to the rear of the flight deck, another in the

starboard side of the nose section and a third from the underside of the fuselage.

Dornier Do17s were used extensively by the Luftwaffe as support to Hitler's armies as they *blitzdrieged* Europe. During the Battle of Britain and "Operation Sealion," Hitler's grand design to take Britain, Do17s were used quite heavily, attacking shipping in the English Channel and hitting various targets in the British Isles.

Dorniers struck at Polish airfields and aircraft factories with high efficiency. After Poland there was a lull in the fighting, the "Phony War" of the winter of 1939–1940, as all sides took a breather and regrouped. During this lull Do17s paid regular visits over France and dropped leaflets and took photographs but released no bombs. Then in May of 1940 the real war was started when the Germans blitzed into Holland, Belgium and Luxembourg. On that day in May the commanding officer of *Groupe de Chasse* 1/5, a part of the famous "Stork Wing," was flying patrol from his airfield at Suippes and he encountered five Do17s over Mourmelon at 12,000 feet. Capitaine Jean Accart engaged the Dorniers and one was shot down and the remainder turned away from the action. Later in the day the French flown Curtiss 75A *Hawks* encountered 21 more Do17s striking at Suippes and their airfield and shot down several and dispatched the rest.

The action portrayed is the *Hawks* engaging the graceful Do17.

ATTACK AIRCRAFT

Junkers Ju 87 *Stuka*

The *Stuka,* like the famous sharkmouth P-40s, is legend in the annals of aviation history. *Stuka* is an abbreviation of the German word *Sturzkampfflugzeug* which when translated means literally "dive-bomber." The design broke onto the world scene in the early 1930s and first tasted combat during the Spanish Civil War in the late '30s, and the airplane was highly effective in that action.

When Hitler unleashed the infamous *blitzkrieg* on Poland the first of September 1939, the ugly, bird-like *Stukas* hurled their fury unmercifully on the Polish defenders. Unopposed the *Stukas* screamed down onto bridge sites, factories and the great city of Warsaw to bring the fear of death wherever they attacked.

After Poland came the conquest of the Low Countries, Holland, Belgium and Luxemborg, and the terrorizing and deadly accurate *Stuka* was again in consort with the lightning fast German army. Fighting again, unopposed in the skies, the *Stukas* added to their terrifying reputation of being the carrier of great fear, death and destruction. It was an honest assessment! The aircraft's vulnerability began to show itself, however, in subsequent actions over France where the Curtiss *Hawk* 75s were able to badly maul unescorted Junkers Ju 87s.

And when the German noose closed on Dunkirk, Luftwaffe *Stukas* were sent in to demolish the shipping that had come to rescue the retreating Allied soldiers. It was here that the beginning of the end for the terrible *Stuka* took place as defending *Hurricane* and *Spitfire* pilots ripped the dive-bombers to pieces.

It was the Battle of Britain that rang down the curtain for the Ju 87; the craft simply could not be defended when up against *Spitfires* and *Hurricanes,* and any German fighter escort advantages were lost the instant the *Stukas* began their dives. The airplane was exceptionally vulnerable right after bomb release, during the pull-up. The resulting *G* forces usually blacked-out the crew and left the airplane defenseless for a moment after the action.

While the Junkers Ju 87 really lost its terror during the Battle of Britain, operational use of the aircraft continued on all fronts right up to the end of hostilities.

Probably one of the most famous of all *Stuka* pilots is Hans Ulrich Rudel, portrayed here as he lines up on yet another Russian tank. It is said that he participated in excess of 2,500 missions and attacked practically every kind of target. Among Rudel's victims were the Russian battleship *Marat,* which he sank with a 2,200 pound bomb placed on target from a 90 degree angle of attack, and more than 500 enemy tanks.

Douglas A-20 *Havoc (Boston)*

Douglas A-20 *Havocs* were the main attack bomber of the Army Air Corps when America entered the war. It was an extremely popular aircraft and ultimately flew combat missions in all theaters of war and in the colors and markings of many of the Allied nations.

The French were early importers of the *Havoc,* the export version labeled DB-7. It was this sleek aircraft that prowled the North African battlegrounds of Tunisia with French crews from Squadron 15 out of Youks-les-Bains. These *Bostons* (as the *Havocs* were called when exported) fought the elite Afrika Korps all over Tunisia. They were everywhere, ripping into German tanks, hugging the hot desert sands as they roared in to mangle General Rommel's well-camouflaged field artillery positions, bursting out of nowhere to disrupt advancing convoys, keeping storage and supply depots in a constant state of confusion and disrepair, and making life miserable for Axis shipping, usually secondary but delightful targets.

Before 1942 was out the U.S. Army Corps 47th Bomb Group arrived from the States with original order Douglas A-20 *Havocs* and took over the 15th's remaining battered DB-7 *Bostons.* From then on the 47th flew support for the U.S. II Corps and hit German targets mainly in central Tunisia, Sfax, Gabes and along the Mareth Line.

In the painting are three 47th Bomb Group *Havocs* attacking shipping running out of Sfax harbor.

Douglas SBD *Dauntless*

While this painting deals strictly with elements of Lieutenant Commander Clarence W. McClusky's eighteen SBDs from the U.S.S. *Enterprise* as they plunge from 17,000 feet to pulverize the heavy carrier *Soryu,* an appreciation of the Battle of Midway would slip by if the reader were not aware of some supporting information. A brief look at the differential in the strength of the opposing forces is both revealing and astonishing.

Yamamoto, the architect of the Japanese adventure at Pearl Harbor, was also the designer of the Midway action. And he was nobody's fool. His strength included eight carriers (four heavy class), eleven battleships, a mixture of twenty-two light and heavy cruisers, more than sixty destroyers and in excess of twenty submarines—a formidable force of warships in anybody's Navy. In addition, at least 700 aircraft consisting of dive-bombers, fighters and torpedo bombers were at his disposal. In their cockpits sat a respectable number of highly seasoned, battle-wise, first-line pilots. And topping it all off was the fact that Yamamoto was the most brilliant naval strategist/tactician produced by Japan in modern times.

Pitted against this vast array of sea power was Rear Admiral Fletcher's relatively inferior force. It consisted of three carriers (one of which, the *Yorktown*, had been hastily repaired at Pearl after sustaining damage during earlier bloodletting in the Coral Sea action), one light cruiser, seven heavy cruisers and fifteen destroyers. And he could put into the air some 230 aircraft from his carriers and cruisers plus an assortment of navy and marine combat aircraft, based on Midway, that totalled slightly under 100. Then there were the seventeen Army Air Corps B-17s from Hawaii and the four B-26 *Marauders* converted into torpedo bombers. Anyway you figured it, the numerical odds definitely favored Yamamoto.

Yamamoto divided his forces into three groups—one to move north and attack the Aleutian Islands off Alaska, the second to attack Midway, while the third, from where he would direct the battle, maneuvered well out of the fighting arena but close enough to move in when the American Navy showed itself.

His intended objective in sending a force north to the Aleutians was to draw out into the open the remnants of the U.S. fleet. It was determined the Americans would surely move to vigorously oppose any offense made against their territory, regardless of its geographical location. Concurrent with this part of the strategy, Admiral Nagumo's 1st Carrier Force would drive hard for Midway. Fortunately, early in the morning of 3 June 1942, a lone PBY spotted Nagumo's force moving toward Midway.

Mid-morning on the 4th of June, McCluskey led his sixty-one aircraft (thirty-seven SBDs) from the U.S.S. *Enterprise* to seek out Nagumo's carriers. A similar force from the *Hornet* was launched. Fletcher, aboard the *Yorktown,* committed thirty-five more to the offensive flight. After great frustration in searching for an elusive enemy, aircraft from the *Hornet* spotted half of Nagumo's carrier force and sped to the attack. Almost immediately torpedo bombers from *Enterprise* and *Yorktown* dashed for the other half that they had spotted. Of the forty-one planes that attacked thirty-five were destroyed by *Zeros* that swooped down from above or by deadly accurate anti-aircraft fire from the carriers and their escorts. Carrier defense against the torpedo planes was so effective that not a single torpedo scored a hit.

The sacrifice, however, was quickly and thoroughly avenged. The carrier's protective screen of *Zeros* that had gone to the deck to maul the torpedo bombers had left the skies clear for the SBDs. *Dauntlesses* from *Hornet* and *Yorktown* selected *Kaga* while McClusky divided his *Dauntlesses* into two formations and hit *Soryu* and *Akagi*. The fourth carrier, *Hiryu,* was far north of the action and was left alone until later in the evening when *Dauntlesses* from *Enterprise* laid four solid hits on her flight deck.

By the morning of 5 June, all four carriers of Nagumo's 1st Carrier Force had been sunk along with several ships of the line. The Japanese Imperial Fleet had suffered its first defeat in 300 years.

Of the thirty-six aircraft depicted in this gallery of paintings, no less than eight types were actively engaged in the fighting at Midway, the PBY *Catalina,* SBD *Dauntless,* F4F *Wildcat,* B-17 *Flying Fortress,* B-26 *Marauder,* TBF *Avenger* and the *Zero.*

Yamamoto's diversionary force in the north yielded nothing for him. It did, however, give the Americans one crash-landed *Zero* in excellent condition. Flight tests of this airplane were instrumental in developing its nemesis, the F6F *Hellcat.*

Douglas A-26 *Invader*

With the Battle of the Bulge behind them, the Allies once again began moving the irregular-shaped battle lines to an increasing shorter length as positions were firmly stabilized, and the noose was drawn tighter and tighter around Nazi Germany. The end of Nazism would come on 7 May 1945, but not before a reluctant enemy made his last ditch stands.

By the 21st day of April, seventeen brutal days and nights before a silence would descend on the larger part of Europe, the once powerful Luftwaffe had seen its Wagnerian "dawn of the gods" fade from existence. Allied airpower struck hard at targets that gave back little opposition. The earlier "Operation Clarion," a concentrated effort of interdiction in which the fire-breathing A-26s inflicted burning waste on the German communication and supply-systems, had been highly effective. Then, "Operation Grenade," which came immediately on the heels of "Clarion," saw the A-26s chewing up German armor in front of the advancing Allied foot soldiers in their drive to the Rhine.

As the noose closed on Germany, the 9th Air Force's mediums, which included the fearsome *Invaders*, intensified their roles of interdiction, always a few steps ahead of our racing armies routing the badly beaten enemy that had once invaded, then subjugated, no less than fifteen nations. As freedom came to these nations, their people swarmed into the streets welcoming and greeting their liberators. And overhead, hurtling to the next concentration of Germans making their last stand were the deadly A-26 *Invaders* of the 9th Air Force.

On the 21st day of April in 1945, the 386th, 391st, and 416th Bomb Groups of the 9th Air Force (all flying the slender A-26s), along with Douglas A-20s of the 410th, flew in excess of 120 interdiction sorties to raze the railroad marshalling yard at Attnanpucheim in Austria. The low flying bombers caused tremendous damage, witnessed no flak at all, and lost not a single aircraft as they also cut the main line from Vienna to southern Germany.

Bottled up in a small pocket of western Germany were the remaining seasoned troops of Germany. They would go nowhere! On the 28th of April, only a few days after the massive interdiction raid on Attnanpucheim, nattily uniformed Germans entered the Allied Headquarters located in the ancient palace of the kings of Naples at Caserta, Italy, and began dickering for surrender terms.

By the 1st of May, only eleven days after the raid on Attnanpucheim, the word would be out that Hitler was dead.

One by one the death dealing rattle of the *Invader's* guns fell silent, and their crews slipped the racy bombers into landing patterns for the last time.

It is noteworthy that while the A-26 only managed a beginner's track record during World War II, it was a good one—so good, in fact, that the craft was resurrected many years later to tear up the Ho Chi Minh Trail wandering down from the north into South Vietnam.

Curtiss SB2C *Helldiver*

The Battle of Leyte Gulf is considered to have commenced on 23 October 1944, and to have ended on the 26th. Because of the sheer numbers of warships and aircraft involved and the sprawling expanse of territory over which the battle took place, the action is justified in being remembered as the greatest sea battle of all times.

As in practically all other Pacific engagements of any magnitude, it was the airplane that made the difference. It was here during the waning hours of battle that the Japanese effectively introduced the *Kamikaze*.

Many months prior to Leyte, the Japanese had developed a master strategy plan code-named Operation *Sho; Sho* meaning Victory. Operation *Sho* considered four offensive actions to be taken by the Allies, one of which, *Sho-1*, concerned the Philippines. In countering an offensive in the Philippines, *Sho-1* would, and did, bring practically all remaining naval and air power of the Japanese into the action; it was a pincer movement from north and south complemented by a diversionary force to draw U.S. carriers and planes away from the main battle site. And so it was that Admiral Halsey, Commander-in-Chief of the 3rd Fleet, was sucked in. He raced north into the Philippine Sea to have it out once and for all with what he considered to be the main attacking force of the Japanese. Halsey took more than sixty warships and 750 aircraft into an engagement against Vice-Admiral Ozawa's seventeen warships carrying a meager thirty aircraft, the decoy force of *Sho-1*.

Needless to say, Halsey's tremendous advantage tore Ozawa's force apart, sinking all four of his carriers, a cruiser plus destroyers, and downing every one of the aircraft. On the 25th of October elements of the 3rd Fleet were only forty miles from Ozawa's fleeing, battered force when orders from Nimitz turned the bulk of Halsey's forces southward to enter the fierce fighting that was raging near Leyte Gulf.

Helldivers played a major role in turning *Operation Sho-1* into a disaster for the enemy. Never again would the once mighty Japanese Imperial Fleet pose a threat to the advancing Allies.

On 26 October it was all over, but not before this *Helldiver* pilot from the U.S.S. *Hancock,* a part of Task Group 38.1 left behind by Halsey to finish off Ozawa, selects one of the scattering remnants of the decoy force as his target and drops down for the kill.

The *Helldiver,* or "Big-tailed Beast" as the aircraft was frequently called, was the last design in a long line of pure dive bombers built by America. Plagued by a host of design changes, the big craft was late in making her appearance with the fleet in the Pacific; however, she replaced the tired but steady *Dauntless* in time for the Battle of the Philippine Sea.

Grumman TBF *Avenger*

Grumman-designed and built TBF *Avengers* entered the Pacific war in early June, 1942, just in time to make their fighting debut in the Battle of Midway.

Assigned to operate from the U.S.S. *Hornet* with Torpedo Squadron 8, TBFs were aboard the carrier with six held in reserve, land-based on Midway Island.

It was these six reserve *Avengers* of Torpedo Squadron 8 that were committed to the fighting on the first day, and it was these *Avenger* pilots who made first contact with the enemy. As they streaked in fast and low for the carriers, swarms of screening *Zeros* dropped down onto them. Breaking through the *Zeros,* the remaining *Avengers* were met by a wall of blistering anti-aircraft fire from the enemy warships. Not only was the flak heavy, deadly, and accurate, but shells exploding in the water ahead of the torpedo planes threw up pillars of water equally as dangerous as the gunfire.

The heroism of Torpedo Squadron 8 is reflected here as we see Lieutenant A.K. Bert Earnest piloting *Avenger* 8-T-1, the only surviving aircraft of Torpedo 8, through withering fire as he hurtles toward his target, the Japanese carrier *Akagi.*

On the second day of battle the remaining aircraft of Torpedo 8 were committed to action from the deck of the *Hornet.* None returned.

Although thoroughly mauled at Midway, the big *Avengers* bounced back to be a continual menace to the Japanese. Among the more prominent victims to fall before the *Avengers* were the two Japanese super-battleships, *Musashi* and *Yamato.* In late October, 1944, *Avenger* pilots caught *Musashi* out of Singapore enroute to participate in the Battle of Leyte Gulf and sent her to the bottom with three solid hits. *Yamato* met a similar fate in April of 1945 as she tried in vain to break up the invasion of Okinawa.

It is interesting to note that *Avengers* went on to play important peace-time roles as civilians after the war. A redesigned bomb bay made the huge plane exceptionally adaptable as a forest fire water bomber machine.

De Havilland *Mosquito*

Launched by British Air Marshal A. T. "Bomber" Harris, the first 1,000 bomber plane raid, dubbed "Operation Millennium," completely overwhelmed the Rhine River city of Cologne, Germany, the night of 30 May 1942. The resulting devastation was but a prelude to what was to become a recurring nightmare for the Germans.

The Air Marshall left no psychological stone unturned. In the early morning hours following his first "Millennium," a new sound descended upon Cologne when four speedy *Mosquitoes* scattered Germans busily engaged in cleaning up the damage from the earlier massive raid.

Eventually "Mossies" became highly specialized in this kind of follow-up operation, moving in quickly behind heavy bombardments with a couple of "block-buster" bombs, then returning home at high speed. Although they were capable of being heavily armed, Mossies usually flew pathfinding missions (spotting the target area for the following heavy bombers) and nuisance raids completely unarmed. With the lightened load, the twin-engined craft were so fast that the best the Luftwaffe could put in the air rarely caught a "Mossie." This tremendously effective aircraft enjoyed the lowest loss rate of any aircraft within Bomber Command.

It is significant that the tactical performance and design of this plane had such an impact on Hitler that it was instrumental in delaying the production of the twin-jet Messerschmitt Me262 for an entire year so it could undergo redesign to parallel more closely the fighter-bomber configuration of the *Mosquito*. This interruption of the Me262 program possibly denied the Luftwaffe a potential aerial warfare superiority that undoubtedly would have altered the course of the war.

In the early morning of 11 April 1944, Wing Commander R. H. Bateson of No. 613 Squadron led six *Mosquitoes* fast and low over Holland. His objective? The five story Gestapo Headquarters across from the Peace Palace at The Hague. Here we see two of those six *Mosquitoes* just after they have neatly bounced a pair of bombs through the front door of the Gestapo building.

Aichi Type 99 D3A1 (*Val*)

This first-line Japanese dive-bomber was first introduced to action during the China fighting. The Aichi Type 99 went on to become a menace to Allied fleet elements in the Indian and Pacific oceans.

Val dive-bomber pilots were among the best in the world, a result of intensive training. Their proficiency in dive-bombing techniques gave them a fantastic "direct hit rate" of 82 percent effectiveness during naval action in the Indian ocean.

Val dive-bombers were in the action all the way. At Pearl Harbor there were 51 Aichi Type 99 aircraft in the first strike group led by Flight Lieutenant Kakuichi Takahashi. These first strike dive-bombers scored heavily on the sleeping U.S. Navy and Army Air Corps. The second assault group entered the Pearl Harbor action at 9:00 A.M. and 80 *Vals* finished off the job started by the earlier strike group.

As the war moved closer to the homeland islands it was the trusty Aichi Type 99 that saw extensive use during the *Kamikaze* actions.

The craft was so versatile that on occasions it doubled as a fighter.

In the action depicted we see a *Val* off the carrier *Soryu* as its pilot lines up on a Pearl Harbor target.

PATROL BOMBERS

Lockheed *Hudson*

What a machine this lovely lady was. War was not her "thing," but the far-sighted British knew a good bomber when they saw one.

A product of Lockheed ingenuity during the middle 1930s, *Hudsons* (as the British would come to call them) were designed to compete with Douglas DC-2/3 type aircraft. Lockheed's plane was shoved into the world's limelight in 1938 when young Howard Hughes circled the globe in one, taking only three days, nineteen hours and eleven minutes.

British aircraft purchasers in this country asked Lockheed to give them a proposal covering a bomber version of the plane because they liked what they had seen of the aircraft's performance, and Lockheed could obviously turn them out in acceptable quantities. Most important, however, was the speed of the big, ruggedly built Lockheed airplane. It was faster than anything the British had in their inventory.

Lockheed's bomber-version proposal was accepted, and deliveries of *Hudsons* began shortly before Hitler's attack on Poland brought France and Britain into the war.

Almost immediately German U-boats began raiding British bound convoys, sinking ships loaded with the badly needed *Hudsons*. The plea went out to begin ferrying them by air across the wild Atlantic, and overnight the losses of *Hudsons* dropped significantly.

Hudsons quickly became the mainstay of the Coastal Command and were rated in popularity by the Britishers as second only to their magnificent Supermarine *Spitfire*.

Every plane has its great moment of glory. For the sturdy *Hudson* it was Dunkirk. Like mother hens guarding their flocks, the *Hudson* pilots gave unceasing support to the bloodied British soldiers who had managed to make it to the beaches at Dunkirk.

More than two hundred German U-boats were sunk by *Hudsons;* countless others were damaged, and one was even captured by a 269th Squadron *Hudson* one August morning in 1941. Damaged to such a degree that it could not submerge, the German gun crews swung into action against the *Hudson* only to be shot to pieces. Each time the U-boat skipper put another gun crew to work, the *Hudson's* guns would down them with blistering return fire. Unable to submerge or fight, the skipper could only give up.

Hudson armament varied, especially as time wore on. In its earlier actions, however, it sported five 30-caliber machine guns, two in the Boulton-Paul top turret (which, incidently was installed in Britain rather than during production stateside), two in the nose section mounted in a fixed position above the bombardier and one belly gun.

The USAAF designation of the *Hudson* was A-28 Attack Bomber while the Navy designated it PBO-1 Patrol Bomber.

Here we look down on a familiar *Hudson* pastime, catching a German U-boat running on the surface. A pair of them from the Coastal Command begin the deadly act of destroying it.

Consolidated PBY *Catalina*

Somewhere in the middle of the Coral Sea a *Catalina* begins cranking down its wingtips preparatory to making a landing in moderate seas. Two more airmen will be saved to fight again.

Tales of the venerable PBY are too bountiful even to begin to relate here. Many action-packed scenes of daring, victory and tragedy remain to be painted about this slow moving, low-flying, long-range flying boat that enjoyed the largest production run of any flying boat in history.

In addition to a formidable array of machine guns, the *Catalina* (named by the British and adopted by the U.S.) could prowl the sea lanes carrying four depth charges, or up to four thousand pounds of bombs or a couple of torpedoes mounted beneath the huge wings.

Strong construction permitted the PBY to absorb great punishment by heavy seas or eager enemy fighter pilots. While plenty of "May Day" victims owe their lives to the air-sea rescue ability of the "Cats" or "Dumbos," as the great craft were fondly called, the enemy also had a healthy respect for its other versatile uses. In May of 1941, an R.A.F. Coastal Command PBY shadowed the mighty German battleship, *Bismarck*, as it broke out of the Baltic. *Catalinas* found the Japanese Midway Invasion force and then later were the first to attack, thereby opening the Battle of Midway. The last to leave Manila Bay and little Corregidor, the island symbol of a national pride, were a pair of PBY *Catalinas*. Coastal Command *Catalinas* posed a continuous threat to German U-boat crews, and more than one U-boat fell prey to the Royal Canadian Air Forces "Cansos," the name given to the PBY by our Allies to the north.

By some estimations more than one hundred PBY's still roam the skies today, a fine testimonial to a great plane who had its maiden flight way back in March of 1935!

SCOUT AND LIAISON AIRCRAFT

Vought-Sikorsky OS2U *Kingfisher*

Kingfishers were in the action all the way, from the attack on Pearl Harbor to the "signing" on the decks of the *Missouri*. In between these two momentous events the rugged OS2U served the fleet in numerous capacities but predominately as "eyes" for the big guns of the battleships and cruisers and for rescue work.

It is the latter that we selected to present the OS2U *Kingfisher*. The people of Yaptown, Yap Island in the Carolines, basked in a pleasant, typical Pacific morning sun on the 27th of July 1944. The big harbor was peaceful; only an occasional whitecap broke the monotony of its surface. Overhead were scattered, fluffy cumulus clouds, cotton-balls hanging listlessly in the rich blue sky. Then, Task Force 58 with its 125 ships of the line (sixteen carriers) struck from fifty miles away. Around mid-day *Avengers* and *Hellcats* of VT-31 and VF-31 from the carrier *Cabot* struck Yaptown from the inland side of the city. The *Avengers*, each armed with four 100 pound general purpose bombs, made their

33

bomb runs from 15,000 feet dropping down over the target then levelling off at 500 feet as they raced out over the harbor.

In mid-run to target a TBM of VT-31 is hit in the main fuel tank by flak. The pilot drops his bombs. The cockpit is burning. Still he holds the big ship at 500 feet until his two crewmen finally clear their aft compartment, which is also burning. Only then does the pilot leave the burning aircraft. Three are burned and down in the harbor. The enemy shore installations begin firing on them, but the shots pass overhead. The guns cannot be depressed low enough to get them, and the three are just beyond the range of smaller weapons. While VF-31 *Hellcats* go to work strafing the shore batteries, a call is flashed to the anti-sub patrol for a *Kingfisher.*

Within minutes a *Kingfisher* from the cruiser *Columbia* was on the scene, its pilot bringing the aircraft in low and straight for the wounded *Avenger* crewmen. Now the shore batteries took him on as a target. The water came alive from exploding shells, still the OS2U came on to the men. After loading the TBM's pilot into the rear seat, his two crewmen sat on the wings and held on. The OS2U pilot would now dash over the water for the pick-up submarine five miles away. On the way out the *Kingfisher* pilot must pass through the entire effective range of the guns on the beach—and it is this section of the action that you see in the painting.

The end of the story is that thanks to a *Kingfisher* and her pilot and radioman-gunner, three more American lives were saved. After the long taxi to the submarine, the two *Avenger*-enlisted crew members were transferred. Then, with the *Avenger* pilot still snuggled down in the rear seat and its usual occupant sitting astride the fuselage just behind him, the OS2U pilot took off for a rendezvous with the *Columbia.*

Stinson L-5 *Sentinel*

What do you do when the mission is to hit Madang during mid-morning and you find them hitting back? On a bad day you might feel that unmistakable shudder ripple through your rugged A-20 and know you have taken a bad hit. The gauges get around to verifying your suspicions. The starboard engine growls into a twisted mess. Flames slice

over the wing, an omnious prelude to a fatal explosion. The entire empennage shakes so violently that you wonder what is holding it all together.

It has all happened so fast that you are a healthy fifteen miles downwind before you know it. Your altitude is slightly above tree-top level. Then, as if God willed it, ahead of you manifests a patch of lovely Kunai grass. Everything is either feathered or shut down. The word is hastily passed to prepare for an unscheduled landing. Flaps down full, nose up, gear up, fingers crossed, you let her down. Perfect! You've temporarily ruined a swath of the plush eight- to twelve-foot-high Kunai grass, but not one additional wrinkle was put in the plane as you sat her down.

Worried? Not particularly! A Jap patrol, having seen the masterful aerial demonstration, might saunter in for depositions and then take you on a nasty hike; or hungry *Zeros, Rufes,* or *Franks* could happen by looking for a helpless statistic, but more likely the beautiful red and white striped tail of an all silver 25th Liason Squadron L-5 *Sentinel* would "mosey" by to assess the situation. It's the Guinea Short Lines at work, dedicated for some mystical reason to pull off the impossible, if only to satisfy themselves.

Getting down is one thing; getting back up is quite another. Like yourself, the L-5 pilots can ease into the high, cushioning Kunai. Getting out is the problem. Everyone goes to work hacking down the Kunai transforming it into a makeshift runway. After an enormous amount of effort, it's time to try. Four-hundred pounds is the L-5's gross load limit according to the unused tech order. You weigh a solid 175. The pilot is a definite 160. Lots to spare! You case your lanky co-pilot and overweight gunner, then almost choke when the L-5 pilot decides on the gunner. Reluctantly you sit in your gunner's lap and look over the L-5 driver's shoulders. Ahead of you is a nasty tree line that rises frightfully high. Way overweight, a short makeshift runway . . . everyone sweats a little. It's now or never. Grinning, the sergeant pilot sets the brakes and revs it up until the wings are about to shake off. Just when the whole thing seems ready to disintegrate, he releases the brakes and the L-5 lurches forward. The trees loom up ahead. Death seems imminent. A miracle occurs as the wheels leave the stubble. The trees are taller than ever. To climb over them is aerodynamically impossible. Without giving the impending catastrophe a second thought, the sergeant pilot makes a squeaky left bank. Completely rigid, you watch as (nobody will believe this) he calmly threads his way through the jungle gaining altitude at every opportunity. Dumbfounded, you suddenly become aware that he has broken out of the tree line and into blue sky. You breathe again and wonder if all those promises to God will be kept.

It is all in a day's work for the unheralded men of the Guinea Short Lines, really the 25th Liason Squadron. They gained the title for getting you from your downed position back to home base in the most direct line; hence, Short Lines. Their feats are legend. Some of their stories are so bizarre that only those airmen involved would believe them.

This selected scene portrays the most common mission/task confronting the 25th. It speaks fairly well for what it was all about.

Messerschmitt Bf 108b *Taifun*

Had Willi Messerschmitt, the great German aircraft designer, been a less dedicated man in his advance thinking, the mainstay of the Luftwaffe, the dreaded Bf 109, would have never existed.

The year was 1934, the middle of a world depression and the rising tide of Nazism. Willi chose this time to unveil the most remarkable aircraft design ever seen in the sports field—the Messerschmitt Bf 108a. The design received immediate and wide acclaim, and offers began to come in from outside Germany.

Quick on the heels of these orders came the feared Gestapo. Willi's simple logic caused second thoughts in certain quarters of the German hierarchy: "In the absence of local support he was forced to design and sell to foreign markets."

Professor Messerschmitt was permitted to enter the 1935 fighter design competition and almost immediately faced the obstacle designed to bankrupt his operations. While all other competitors were to receive German-made engines (compliments of the government), *none* were available to Messerschmitt.

Willi's design was not only simple but extremely functional and, as it turned out, flexible. Going with what he had in the 108 design, he reduced the girth for four passengers down to a sleek one that would accommodate a lone pilot. This, in turn, allowed a lengthening of the fuselage which was necessary to house a more powerful engine, accessories, armament and larger fuel tanks. However, the 108 nose, wings, and tail section remained remarkably the same in appearance in the new 109 configuration. And Messerschmitt solved the riddle of the engine problem by incorporating a less powerful but satisfactory British Rolls-Royce Kestrel V engine, the same powerplant that would be pitted against the Bf 109 in later years inside England's Supermarine *Spitfires*.

The highly modified 108 and its British powerplant lost out to Heinkel in the competition trials. However, Willi's 109 gave good enough account of itself that the fledgling Messerschmitt organization received an order for ten production models. The rest is history. At last count in excess of 33,000 Bf 109s were manufactured in a multitude of variants and by many European firms, the longest production run of any aircraft in history. And it all began with the radical design of the Bf 108a.

During the war years Messerschmitt's Bf 108bs were built in limited quantities to serve as liaison and courier duty aircraft. Such is the scene depicted here; a *Taifun* pilot gives his passengers a look at Castle Neuschwanstein as he threads his way through the Bavarian Alps enroute to the Black Forest from Salzburg.

Missing Man Formation

Long recognized within the flying community as the highest honor that can be bestowed upon those airmen who made the supreme sacrifice in the defense of their country is the solemn heart-rending tribute known by airmen the world over as the *Missing Man Formation*.

It signifies that a pilot or crew for reasons beyond their control cannot fly and fill the formation. The *Missing Man Formation* is usually executed at low level, about sunset, and is composed of either four or five aircraft that do not all have to be of the same type.

For this formation and scene five aircraft were used, all of which are F4F *Wildcats* of VMF-211, the Marine Fighter Squadron destined to fly into history during their courageous stand at lonely Wake Island in late 1941. It was a desperate but futile battle of which there would be no surviving aircraft. Helplessly, the U.S. Navy and A.A.F. sat by and died a thousand deaths as the tiny fortress fought back with all the heart and soul possessed by men. Absolutely vulnerable to the guns of the Japanese warships and the overwhelming numbers of aircraft types that would be launched against them, the Marines of Wake Island preferred to fight to the end rather than take the easy way out and surrender without putting up a fight. Major Devereaux commanded all of Wake, its 478 tough marines, their few batteries of five and three inchers plus twelve *Wildcats,* four of which were operational most of the time.

Nonetheless, they held the line from 7 December 1941 to 22 December when the Japanese landed an overpowering force and shot down the last two remaining F4Fs but not before the pilots had sunk an enemy destroyer.

The loss of these courageous fighting Marines and the entire force of VMF-211 exemplifies the discipline and determination of the U. S. Marine wherever he is found. It is only fitting that the *Missing Man Formation* saluting all missing airmen should be composed of VMF-211 *Wildcats* off Wake Island.

Epilogue

So, where did they all go; the aircraft and the warriors, those who fought *and* those who built? It is a question, a subject, that is very rarely mentioned even in a passing way. We simply never give their final disposition another thought. It was a bad dream that is now behind us and we are prone to forget such things as quickly as possible . . . it is impossible to "recall" pain.

We all remember it was a fight to the finish . . . no compromising. And that is decidedly how it was concluded. Part and parcel of the surrender terms was the total destruction of the enemies' capacity to build instruments of war. This included the demolition of *all* flight items save a precious few destined for engineering study. The back of the remnants of the Axis aerial might was summarily broken . . . forever.

This demolition was not limited to the Axis powers alone, however. Ramps of new A-26 *Invaders* were rendered useless by dynamite on their European bases. Thousands of B-24 *Liberators* were destroyed on tiny Biak Island, New Guinea, and left for the jungle to consume them.

And so it went around the world. The order of the day seemed to be to render useless all but a precious few, guardians of the peace. At home the destruction was even more pronounced. Here we had the great smelters, the fiery furnaces that could in the twinkling of an eye reduce a spanking brand new B-17 or sleek P-51 into a batch of gleaming aluminum ingots. Arizona in particular became the national graveyard for all these majestic, freedom-saving aircraft. The furnaces operated at full capacity to literally keep Arizona from becoming polluted with aging aircraft, not quite unlike beaches full of beer cans; a poor analogy. Strangely enough the process continues until this very day. True, we have up-graded our methods of destruction and become more sophisticated in the eradication process. Nonetheless, you can visit Davis Monthan AFB, Arizona, and see the once proud fleet of B-58 *Hustlers* neatly lined up in a row silently awaiting their turn to be chopped to pieces and crammed into the careless furnaces.

Just as fast as we put behind us the last generation of aircraft (whether they fought in war or were deterrents in peace, is of no consequence) we also tend to abandon those who flew them, or crewed aboard them, or supported them on the ground, and those who designed and built them. Therein lies the real tragedy.

But fortunately for us the world is a big place and the bureaucratic process has its shortcomings in *some* instances. As a result many of these history making planes managed to get lost in the shuffle. Unbelievably, some were even offered for sale as war surplus material. Between the "mavericks" and those sold, a fair assemblage of aviation's history from 1939 through 1945 (and later, too) has been salvaged for personal use of one type or another, and/or posterity. While it is rare, it is not altogether uncommon to round the corner in some relatively unknown city or town and suddenly come face-to-face with a deadly looking World War II fighter usually cocked at an angle atop a stone or iron pedestal. Others lie unnoticed in falling down barns or in unkept sections of fields . . . awaiting to be found. In recent years a sense of urgency has pervaded certain segments of our flying community that is sensitive to the need of preserving aircraft that flew in that era of our aviation history.

Because of their fierce dedication it is now possible for people from the world over to see and marvel at historical items that would have certainly perished forever had it not been for this handful of men. Museums now exist all over the world. None are complete, but each presents at least one aircraft not found in any of the other museums. The analogy would be to libraries . . . some are limited, but good, while others are amazingly well rounded out in their inventory. Some deal strictly with Naval aircraft, some are faithful to USAF types while others concern themselves with either foreign built planes or a hodgepodge of them all.

But one thing is certain. With the passing of each day attrition rears its ugly head to take not only one or more of the precious treasures of flying history but also those souls who put them in the air and those numberless ones who rolled them off production line. Eventually all of the people will be gone . . . not so with the aircraft to which they related. Archivists, museum curators and hardheaded individuals will see to it that most of these planes will survive through the ages.

Thousands of excellent books have been written about these great aircraft. From them, an in-depth knowledge is gained. Some of these writings will survive for generations to come. Countless photographs taken during these years exist. Most, however, were static or were candid in nature, and many were poor in quality or point of view. During the fast pace of action it was a very unwise person who attempted to capture the event on film. And if he did the result was invariably poor.

The artist closes the gap and focuses attention on the high-point or interesting feature of an aircraft type or a particular action. Working with pilot's de-briefing reports, good to bad unit histories, and most important . . . effecting face-to-face confrontations with one or more persons directly involved, a true historical event can be visually recorded and supported with the complete story. Done properly, such an endeavor can be perpetuated indefinitely.

This book has been designed to last for generations. The pendulum swings from doves to hawks. As a short-range consideration this book can easily fall within either camp, despised by the one, eagerly sought by the other. In the long pull it will be accepted for the original intention, which has been to add to the historical accounting of that great conflict. In the end this Epilogue should fall within the same frame of reference as all other accounts of our heritage and, in part, some of that of both our allies and our enemies.

VOLUME II

Acknowledgments

Due to circumstances beyond my control, the record of those who provided input to this effort has been lost, if not destroyed. This is regrettable because it was a sizeable list, too lengthy for me to remember in its entirety, and I did want to acknowledge everyone on it personally.

I do, however, recall some who contributed: The U.S. Navy for information on the *Devastator;* Jun Amper, in the Philippines, for data on the P-26 *Peashooter;* a Californian—I think—for furnishing the *Kingcobra* kit when none could be found; the person who furnished the kit correction data on the *Kingcobra* and Il-2 *Shturmoviki;* a Texan—I think you live in Houston—who sent photos taken in Russia that showed how they modified the *Kingcobra* to add more firepower on the upper forward fuselage; a tank modeler from Fort Worth, who supplied the tank used in the Hs 129 scene; a German, who prefers to remain anonymous, for information on the He 219 *Uhu;* all personnel at the Royal Air Force Museum for their generosities to Les while he was there; the Air Museum in West Germany; the Confederate Air Force; and, finally, three persons who made major contributions, and I will never forget their names: William N. Hess of Houston, Texas; Dr. Stanley M. Ulanoff of Roslyn, New York; and Christopher Shores of London, England.

Of course, there were many others . . . and I thank each and every one of you.

Foreword

Here, put into vivid colors and placed on canvas by the deft brush of Tony Weddel, are scenes of World War II aerial combat—combat that brings back memories slowly fading into oblivion for many of us who took part in it.

For those too young to remember, or yet unborn, the paintings are priceless portrayals of engagements fought in a bitter conflict. Through extensive research, attention to detail of battle engagement and airborne equipment, and participant or witness verification, the paintings and their story lines become historical documents that will be cherished more and more as the World War II years fade from living memory.

Though I have personally flown, inspected, or seen most of the aircraft portrayed in this book, my review of the contents found my memory slipping on many of the details. Such things as the color schemes of the Fw 190's and Me 109's of Oberst Josef "Pips" Priller's *Jagdgeschwader,* Number 26, "Schlageter." But well remembered are the flaming spouts that appeared from the forward cowling and leading edges of their wings as mass head-on attacks of these vicious "black-crossed" fighters bore down on us.

They took their toll on us in the early days of 1943. Likewise, by D-Day in June of 1944, we'd clawed and scratched through several score or more of their evasive machines and better pilots.

At the present time, Group Captain Robert Stanford-Tuck, of R.A.F. Fighter fame, and the earlier Kommandant of *Jagdgeschwader* 26, Oberst Adolph Galland, couldn't give you the details of their oppo-

nents' unit colorations or markings either. In an *augenblick* (blink of an eye), as the Germans say, each could provide a verbal schematic of his opponent's fighting machine, but only as a hazy recollection imprinted in the back of his mind.

Offhand, the various performances of opponents' aircraft have also slipped away. It would take a review of old documents to sharpen the gray matter. Thank goodness someone has recorded these details for posterity.

It would be remiss not to recognize the role of the stalwart pilots and crews of the "Big Friends" that carried the payloads of destruction to the enemy heartland, whether they flew the B-17's or B-24's in daylight raids, or rose to fly east during the hours of darkness in R.A.F. Bomber Command's Halifaxes, Lancasters, or Wellingtons. From our fighter bases located on the East Anglia Coast of England, we often heard the early morning hours of assembly of aerial armadas of the Eighth Air Force. Before first light, the continuous drone of their motors drowned the crow of the rooster. From October until late March the weather over the British Isles became dismally worse. By Christmas the "pea soup" fogs, so notorious in London, settled with an ominous presence over everyone and everything. Blind flying became a hazard in itself.

To marshal a thousand or more bombers into combat wings of fifty-plus four-engine bombers was mind shattering. Between clouds, between layers, or above clouds, the show pressed on. Once a string of combat formations gained altitude and established a course, the aerial array stretched 150 to 200 miles. A spectacular display, to say the least. A still more breathtaking scene occurred when atmospheric conditions at 20,000 feet or more became conducive to the formation of condensation trails. The sky became filled with man-made stratus. For those in the cockpits, it was a harrowing day just to cross into enemy territory.

Adrenalin stimulated faster heartbeats from this point on. Heavy, black flak bursts appeared ahead. Those of us in fighters "ducked" around these menacing "steel hailstorms."

Enemy fighters could be anticipated. They could come in waves, in small gaggles, in coordinated strings from head-on, or in fleeting individual attacks from above. Whether the enemy aircraft concentrated attention on a lead "box of bombs" or moved back to launch attacks upon following combat wings depended upon many a factor: i.e., tactics, directions from ground controllers, cloud cover, visibility, bomber stragglers, escorting fighter protection, etc.

A concerted, concentrated attack on one combat wing was most dreaded. Here the attack overwhelmed the mutual protective coverage of turret gunners. Such coordinated attacks literally blew formations apart as "Big Friends" went down with burning engines, shot-out cockpits, damaged controls, exploding gasoline tanks, and a multitude of other unlisted disasters. The objective of an ideal fighter attack was to launch from the sun with complete surprise into a violent head-on attack at very high speeds. The results were devastating and nerve shattering.

Once a bomber formation began losing members, the attention of the opponent concentrated on the lonely singles that struggled to keep airborne. It was bullet for bullet. Hammer, hammer, hammer! A test of the highest degree of courage.

The Luftwaffe pilot held no position of envy, either. It was necessary for him to concentrate on one target, fly through a hail of bullets, and escape to a safe recovery. Likely as not, this sequence of events had to be repeated several times on the same mission. Not an enviable role.

Having spoken with Luftwaffe Ace George P. Eder, I learned that he destroyed 36 four-engine bombers in his overall credit of 78 aerial victories. Eder was shot down seven times in the course of his flying career, winding up in the hospital with a .50-caliber bullet through his stomach. He still returned to fly again, in combat!

At times, a leisurely, late afternoon jaunt along the perimeter track or to the maintenance hangar cleared the mind for contemplation of the day's work. But the rumble of hundreds of motors could be heard in the surrounding countryside as day's light ebbed. The air battle over Europe seldom ceased whether night or day. The R.A.F. Bomber Command awakened under cover of darkness to carry on the "around the clock" aerial assault.

First off were the Pathfinder forces who dropped flares to light up and designate targets. These swift, two-engine *Mosquitos* swept low over our base as they established climbing speed over the North Sea to altitude and the enemy coast. Further to the east lay a somber vastness of foreboding darkness. A blackout prevailed over the continent, but the complex Luftwaffe system of "Wildschwein" stood on the alert. Thousands of crews manned radar. Night fighter pilots waited their call. Hundreds of flak and searchlight teams readied their stations.

By darkness, heavily laden *Halifaxes, Lancasters,* and *Wellingtons* arose from Midland air bases. Except for moonlit nights, when an occasional outline could be discerned, only the rumble of laboring engines was to be heard as the main force passed overhead. With navigation lights turned off, and exhaust manifolds shielded, only the rhythm of synchronized motors prevailed. A thousand or more R.A.F. bombers, flying individually at different altitudes, took over an hour to pass. The stream seemed endless.

Late in the war, as a P.O.W. passing through Berlin, I experienced the devastation wrought by this awesome aerial force. The Berlin sky first opened up with a barrage of anti-aircraft fire that exceeded the imagination. In contrast to daylight, when heavy flak appears as black puffs against the pale blue sky, the explosions of an 88mm shell in the blackness of night comes with a startling flash, followed by a deep rumble. When hundreds of guns are in action, the earth shakes. Shell fragments descending to earth again clatter on rooftops, streets, and sidewalks.

Almost simultaneously, the eerie blackness is broken by target marking flares dropped by the errant *Mosquito* Pathfinder. Dangling from parachutes, the flares light up the surroundings as they slowly descend. A massive explosion follows, which grows in intensity as the big "heavies" unload their destruction. Fires break out; searchlights sweep the sky; the flak barrage continues.

I am escorted into a concrete bunker by my guards, where I crouch as the incessant crash of bombs continues into the night. There is little conversation among those assembled. As an occasional near miss explodes, everyone flinches. A sense of utter helplessness prevails.

How can one describe the actual scene in words alone? Bavousett and his team of artists have caught, with definitive authenticity, instances of rare exposures of the machines and men who clashed so violently. Through these paintings and the synopsized stories which they portray, we are provided glimpses of an era during the late 1930s and early 1940s when people struggled to the death for air supremacy to preserve their respective ways of life. A record and a reminder for future generations.

COLONEL HUB ZEMKE, U.S.A.F. (RETIRED)
Commander, 56th Fighter and 479th
Fighter Groups during portions of WW II

Introduction

Even though I was a deep water sailor during World War II, my first love was combat aircraft. Like legions of other youngsters, I had amassed a great collection of the old solid balsa models and read every printed word about airplanes that could be found. Then, after the war, we all busied ourselves with building futures: The war was now something to talk about occasionally, and the models were set aside. For me the excitement of men and machines tangling in a death struggle in the sky began to fade from memory. Eventually, the whole scene disappeared; it was as though none of it had ever happened. I had forgotten.

Then, in the late 1960's, I had occasion to travel with a close friend to Harlingen, Texas. Dick was going there to be installed as a Colonel in the Confederate Air Force . . . whatever that was! In the back of his truck were two metal detectors, the prime reason for my wanting to tag along. After Dick had done this thing with the CAF, we were going to Padre Island and search for the lost Spanish treasures buried under the sandy beaches. Now *that* sounded like a whale of a lot of fun.

But the glamour of the adventure of searching for buried treasure was totally overwhelmed when I came face-to-face with the realities of the mission of the Confederate Air Force. The place was alive with gregarious Colonels and their ladies, and I was seeing the grandest collection of fighting machines in the whole world. Moreover, *all* of them were in flying condition. I was awe-stricken . . . speechless. Somebody had cared enough to remember! I *had* to be a part of this assemblage of fine men, all of whom shared at least one common bond: A dedication to the preservation, in flying condition, of at least one of all the more prominent combat aircraft types that had fought in the skies during the 1939–1945 era! Unquestionably, this was a mammoth and awesome task.

This experience gave instant rebirth to my dormant love for WW II combat aircraft. I was neither a pilot nor a mechanic. My value as a crew member was doubtful. But I was an artist and I could write. I began boning up on the old warplanes. I was seeing the real McCoy in the CAF hangars and out on the apron. Unconsciously, a thought was forming in my mind.

In poring through boxes and shelves filled with books covering the great WW II air war, I discovered a near total absence of pictures taken during the combat. There was an abundance of verbal descriptions, the tactics and strategies and all that, but no meaningful photographs of the duelling. The thought became a personal mission for me. If the good Colonels of the CAF could find and restore to flying condition these wonderful combat aircraft, then I could cause them to exist on canvas and depict them in raging battles, the very purpose for which they had been designed and built. By creating combat scenes, I would help these Colonels make the world remember those great pilots from all nations and the mighty machines they flew.

Accordingly, in late 1971, I began the effort that would ultimately emerge as "The Valiant Clan," a collection of 36 canvases that at the time included all of the types of aircraft in the CAF fleet that had seen combat. Because I had searched out stories of battles in which these aircraft participated, the elements for a book were present. In 1976 this book was released under the title *World War II Aircraft In Combat,* and it promptly won three awards and became a book club selection.

I began receiving many letters of praise for this work, and nearly all of them asked for more, encouraging me to continue and give similar coverage to other famous combat aircraft. Because of these words of encouragement, and my hopeless love for planes (I admit that I'm hooked!), I pressed on. This companion to the original volume is the result.

The letters I received after the original book was released often posed the question: "Why did you select these particular aircraft?" The answer is, I was essentially tracking those in the CAF fleet. When I decided to commit myself to this companion book, I moved to get as many air enthusiasts involved in the effort as possible. By this time I was selling large lithos of some of the scenes to members of my world-wide "The Valiant Clan Print Club." Using that sizeable mailing list, a questionnaire was developed, asking everyone to name five combat aircraft they would like to see put on canvas, and no fair naming any that had been covered previously. The response was staggering, and, after tallying up everything, it quickly developed that four main groupings existed: German markings, British markings, Japanese markings, and a mixture of markings from other nations.

From these four groups I took the 10 aircraft most often named (adding two of my own choice) and began the research for actions in which each was involved. From this process an individual action was isolated, and from that came the particular configurations, colors, markings and, whenever possible, the background of the scene. We have tried to make them as technically and historically accurate as time and the record would permit. The one exception is the Me-163 *Komet* scene. In Jeff Ethell's book on the *Komet,* he describes this action in great detail, and a reproduction of renowned aviation artist Keith Ferris's painting depicting the "unusual" moment in the action adorns the cover. Both are the result of extensive, in-depth research and as such must be regarded as highly accurate. So, rather than duplicate it, we have fudged to show you these two combatants from the other side.

The one lingering question remains, *"What about the Japanese?"* Well, researching the Rising Sun is a most difficult task. It appears that the historians focused their attention on just about everyone except the Japanese. And while I'm probably very wrong, it seems to me the Japanese have taken the position of "Let's forget the whole thing." Precious little information exists on those planes and their pilots; if it does exist, then it has thus far escaped me. The original plan was to include 12 Japanese planes in this book and leave out the 12 mixed markings. I was making good headway in regard to meaningful stories and the technical information necessary to create excellent scenes when, unfortunately, I was *hors de combat,* so to speak, from mid-1977 to 1979. This put the Japanese on a back burner for more than two years because it was impossible for me then to perform research of this nature. The great Japanese warplanes, and others, will be covered in my next effort.

What you are about to read and see on the following pages took some four years to complete. The planes are presented to you in the order in which the canvases were created: First the British, then the mixed nationalities, to completion with the Germans. It is hoped the stories and their accompanying scenes will fire your imagination, and cause you to glance back in time to ponder on and remember this great era of our aviation history. It is a legacy passed on to us by those who fought for what they believed was right. In time they will all be gone, taking with them their first-hand accounts of how it was; the dwindling supply of the actual planes will surely follow, to leave only the written word and a few remarkable paintings.

My heartfelt thanks go out to all of the Colonels of the Confederate Air Force, for were it not for these super-dedicated men, not a single scene would have come from me.

GLENN B. BAVOUSETT

THE BRITISH GALLERY

True justice for the indomitable British Empire and Commonwealth and Free European pilots and crews, and their aircraft, cannot begin to be exposed by the following twelve stories and their accompanying scenes. At best this gallery only cracks the door to permit a brief glimpse back into a time when these fearless people rose to the occasion, to truly cause ". . . their finest hour," and this applies across the board and not just to that part of the whole—the Battle of Britain—that triggered Churchill to utter the words.

Several of the combat aircraft covered in the following pages were on-line and ready to fight from day one of the war, and they went on to continue fighting until the last day of the war. Others were born later in the unfolding drama, and some of these would continue to bare their teeth and fight in another place at another time.

Like a good wine, the taste of these aircraft lingers on after the glass is emptied. A picture is still just a picture and, as such, cannot raise the hackle of hair on the neck and back like when coming face-to-face with one of these machines that dedicated people have seen fit to preserve. There is a difference. Quite a number of combat aircraft are on public display in various museums in England and the United States. You are encouraged to see them and feel the thrill of your life as you stand there imagining the smells and sounds of war; the electrifying experience of man and machine locked in mortal combat with another man and his machine. It is difficult not to imagine yourself at the controls, roaring down on the Möhne, hot after Wick, driving a torpedo into the *Littorio*, kicking on rudder to put your guns on a *Falco*, diving down to absolutely pulverize a locomotive. Yes, it is still possible to see yourself ease a wing-tip into position to flip a doodle-bug out of control. That, my friend, is the hawk within you rising to the surface . . . the same hawkishness that put those airmen—men just like you—in these planes and made them do what they had to do. For some it is a nostalgic trip back in time, for others a letdown from being born too late, but for all it is a chance to pause and reflect, ponder and remember, for what goes 'round does come back.

The canvases of these twelve history-rich scenes are now a part of the private collection of Douglas Champlin and can be seen at Falcon Field Fighters Air Museum located in Mesa, Arizona.

And now we present to you *Battlers Courageous,* the name given to The British Gallery by a fan of our work.

Westland *Lysander*

" 'Lizzie'—the foot-soldier's friend!" This catch-phrase aptly describes the intended use of this versatile aircraft.

The *Lysander* was designed to conform to outdated specifications developed for army cooperation airplanes, a concept similar to the U.S.A.A.C.'s equally outdated Observation category of aircraft used for front line tasks. In many respects the *Lysander* was very similar to its U.S. and German counterparts—the Curtiss O-52 *Owl* and the Henschel Hs 126.

Operating from rough forward area airstrips, the design criteria called for a multitude of jobs to be performed by the *Lysander.* S.T.O.L. capabilities were high on the list, and Westland engineers satisfied this with the long-span, high-lift wing of unusual plan form that became one of "Lizzie's" most pronounced physical features. As an army cooperation airplane, the *Lysander* was to undertake tactical reconnaissance, direct artillery fire, drop messages, carry urgent supplies to the front line, and carry out limited ground attack duties. For these latter purposes, a .303in Browning machine gun was fixed on each of the heavily spatted wheel coverings—another distinguishing feature that clearly identified the plane to be a "Lizzie." Additionally, outrigger racks—thin, stubby wings—were fitted to the wheel coverings, and these were used to carry supply containers or small bombs. Further, 40-pound bombs could be carried in a small bomb rack mounted beneath the rear fuselage. Lizzie was not without some teeth!

Seven squadrons of *Lysanders* were available to the R.A.F. in September of 1939. Four of these squadrons accompanied the R.A.F.'s Air Component of the British Expeditionary Force when it went to France. These aircraft formed 50 (Army Cooperation) Wing which was there to provide support for the ground forces. During the uneventful winter of 1939–40, these four squadrons of *Lysanders* were joined by a fifth, and all of them saw their first action in the opening days of the *blitzkrieg* in May of 1940.

Usually, only a single *Lysander* went up on a tactical reconnaissance flight, but sometimes they would operate in pairs. Regardless of their number, they frequently fell afoul of the Luftwaffe fighters for which they were usually "easy meat." *Lysanders* had their moments of glory, however, when they managed to outfight their attackers. This didn't happen too often, but it did happen!

Sixteen aircraft from 16 Squadron effected the biggest concerted raid by *Lysanders* on record when they attacked a German convoy near Cambrai on May the 15th.

On May 22nd, Flying Officer A.F. Doidge and his gunner, Leading Aircraftsman Webborn of 2 Squadron, were en route to the Merville area to effect a recon mission when they came across a Ju 87B *Stuka* dive-bomber. The enemy was promptly engaged in combat and Webborn made the kill. Doidge then looked ahead and below and saw a plane of his opposite number—a Henschel Hs 126! Doidge immediately dived to attack, and with a burst from his front guns sent the Henschel down in flames to crash and burn.

A more typical *Lysander* situation for this period is depicted in our scene. The date is May the 17th and the time of day is early morning. Pilot Officer C.H. Dearden, also of 2 Squadron, is at the controls and he's in the Cambrai area when he's jumped by no less than nine of those ferocious Messerschmitt Bf 109 fighter planes. Dearden began taking all of the evasive actions possible with his highly maneuverable *Lysander*, and while none of the German pilots managed to send him down in flames, they eventually inflicted enough damage to force him to land. The badly damaged *Lysander* was later burned on the ground to prevent it from falling into German hands.

The last *Lysanders* were evacuated from France on May the 22nd. Then they operated over Amiens, Boulogne, Calais, and Dunkirk, providing support as required to the retreating Allied army. When operations ceased for these aircraft on June the 8th, the five *Lysander* squadrons had lost a total of 30 aircraft.

Lysander crews also saw limited action over the North African deserts, Eritrea, Greece and Madagascar, and against the Japanese in Burma. Her pilots also flew anti-terrorist missions in Palestine and on the North-West frontier of India. Later in the war, "Lizzie" flew out of England and Burma on clandestine night operations to land secret agents in enemy-held territory.

"Lizzie" was a fine and tough Lady!

Vickers *Wellington*

More Vickers *Wellingtons* were built during World War II than any other British bomber type: 11,000 of them rolled off the production lines. Indeed, only three other bombers were built in greater quantities: the B-17 *Flying Fortress*, B-24 *Liberator*, and Junkers Ju 88. The *Wellington* was destined to remain in service from the first to the last days of the War.

In September of 1939, the "Wimpy" was the R.A.F. Bomber Command's best heavy bomber. The craft was built on Dr. Barnes Wallis's geodetic construction principle which featured a "basket-work" structure of criss-cross light metal framers which were then covered with fabric. This principle required no internal bracing, spars, or stringers, and offered lightweight construction with considerable strength. Since each section of the aircraft was self-supporting, it had the ability to absorb tremendous punishment without crumbling. However, the necessity of cladding the outer surface with fabric, rather than with stressed aluminum, limited its development for flight at higher speeds. Additional speed was the trade-off for being able to take plenty of punishment.

In the early stage of the developing war, it was the consensus of considered opinions of aerial warfare strategists and tacticians that tight formations of heavily armed bombers would be able to penetrate enemy territory without needing any fighter escort, relying instead on the mutual support of their guns for protection against attacking fighter planes. This theory is similar to that espoused by the U.S.A.A.F. when it entered the European war in 1942.

Accordingly, the *Wellington* 1A, which became operational shortly after the hostilities commenced, featured twin machine-gun power turrets in the nose and tail and a retractable "dustbin" turret in the lower fuselage. This "undergunning" would quickly manifest itself for all to see when mutual support failed to be an effective defense against an onslaught by enemy fighters.

A few of these ineffective mutual support missions was all it took for Bomber Command to shift from daylight to night bombing missions. The month of December in 1939 forecast the need to change.

Several daylight raids by *Wellington* formations were launched against units of the German Fleet in the Wilhelmshaven area located on the northern coast of Germany, where the Baltic meets the North Sea. On two occasions the *Wellingtons* suffered bad losses to the guns of German fighters, and then, on the 18th of December, came a colossal disaster.

Twenty-four *Wellingtons* of 9, 37, and 149 Squadrons approached Heligoland on the third daylight raid of the month. A strong force of Messerschmitt Bf 109's and Bf 110's of the composite *Geschwader Schumacher* had been gathered for defense of the area, to handle the very threat that was now approaching. Fully alerted to the menace, the Germans rose from their airfields to intercept the *Wellingtons*. The resulting fight came to be known as "The Battle of the Heligoland Bight" (the subject for our painting) and produced the disaster in which 12 *Wellingtons* were shot down and three others so badly damaged that one ditched during the return flight and the other two crashed when landing.

Further losses in daylight raids over Norway during April of 1940 hammered home the lesson and, thereafter, Bomber Command operated only by night, except in unusual circumstances. While a failure during the daylight hours, the *Wellington* gained great success as a night bomber.

"Wimpy" went on to operate in practically every theater of war in which the illustrious R.A.F. was employed: Over Germany, the African desert, Italy, Burma, and other areas, this great bomber provided the main striking force until at least 1943.

Wellington was the last twin-engined type to be replaced by the newer four-engined "heavies," and it soldiered on much longer than did her other twin-engined compatriots. The craft was also widely used as a torpedo-bomber, an anti-submarine patrol aircraft and, to a lesser extent, for night photo-recon, aerial minesweeping, and transport duties. While never exported, the *Wellington* did serve with the Free European air force within the R.A.F., and was flown by Poles and Czechs.

Boulton-Paul *Defiant*

The *Defiant* was unique inasmuch as it remained the only single-engine, two seat, turret-mounted fighter plane to see action during World War II.

This aircraft was powered by the same 1,000 h.p. Rolls-Royce Merlin engine that was used in the more famous *Spitfire* and *Hurricane* fighters, and the resulting underpower produced a lower all-round performance in regard to combat duties. This was due to the aircraft's greater size and weight when compared to that of the *Spit* and *Hurricane*. The additional size and weight was, of course, brought about by the inclusion of the four-gun power turret and the gunner who operated it.

The *Defiant* was designed specifically to be a bomber destroyer. The theory was that in passing through a bomber formation the turret gunner would be able to continue the attack from many different angles and give a greater duration of fire than a conventional, single-seat fighter which could only fire forward, usually at only one target, during the pass. In this regard the loss of power was not considered to be important. In practice, however, things didn't work out as planned, because the Germans didn't cooperate. They sent fighters along with the bomber formations, and the underpowered *Defiants* became "duck soup" for these pilots. Moreover, once the bombers began to feature heavy armor protection, the battery of four .303in Brownings were no longer particularly effective.

Defiants were first committed to action during the Dunkirk evacuation in May of 1940. Two squadrons of the aircraft were operated as day fighters, and in this role their service was both brief and bloody. *Defiants* were not, however, totally ineffective in combat. Under certain conditions the plane and its crew constituted an awesome adversary that was tough to cope with.

When 264 Squadron made its debut over Dunkirk, its surprise value allowed a few successes to be achieved, for the German fighter pilots mistakenly assumed the *Defiants* were *Hurricanes* and, naturally, employed the classic attack from above and behind . . . and were promptly greeted with a wall of blistering fire from the fully alerted turret gunners! But this type of attack was quickly changed once the

51

Germans were wise to the *Defiant's* aft-firing capability. Now they attacked from below or head-on, where the gunner could not train the turret, and the slaughter of the underpowered *Defiants* was fearful.

The *Defiant's* moment of glory came in late May of 1940. Early in the day of May the 29th, 264 Squadron mauled a formation of unsuspecting Messerschmitt Bf 110 *Zerstörers*. Later that same day the squadron got in amongst a host of slow and lightly armed Ju 87 *Stukas* and cut them to pieces (our scene). Total claims for the day's action reached 37, but actual Luftwaffe losses were substantially less than this inflated number. What with the gunners whirling about in their turrets, many firing at the same targets, and the pilots looking but not firing themselves, the situation was fraught with opportunities for overclaiming, to a degree approaching that of the gunners in bomber formations—and those gunners were notorious when it came to overclaiming!

Soon after the opening of the Battle of Britain, the *Defiant* was released as a day fighter and relegated to the role of a night fighter. More squadrons were formed for this role, but the *Defiant* had no radar and could not be adequately fitted with it. As a result, the *Defiant,* as a night fighter, became nothing more than a stop-gap, and enjoyed few successes in this role.

Fairey *Swordfish*

In 1940, the Italian Fleet posed a massive threat to the British Royal Navy in the Mediterranean, and to the security of British forces in Egypt, Malta, and Cyprus. By November 1940, however, the main Italian battle fleet still had not been met at sea after nearly six months of war, and the British Mediterranean Fleet was doing much as it wished, with its carrier-borne aircraft attacking Italian bases in the Aegean and in Libya, while its surface ships escorted munitions convoys to Malta and Alexandria and raided Italian convoys.

The September arrival of the new armored Fleet carrier, H.M.S. *Illustrious,* allowed a strike—Operation Judgement—to be planned against the main Italian Fleet based at Taranto in southern Italy. Operation Judgement was to be a two carrier operation, but at the last

moment *Eagle* became unserviceable, due to her fuel lines being damaged by concussion from frequent near-misses by marauding Italian bombers.

Then in early November came news that all six Italian battleships were in the harbor—the attack had to be made at once! *Illustrious's* striking force was not large: She carried eighteen Fairey *Swordfish* torpedo-bombers of the 815 and 819 Squadrons, with the fifteen Fairey *Fulmar* fighters of the 806 Squadron for defense. Four more *Swordfish* and two Sea Gladiator fighters from *Eagle's* 813 Squadron were flown aboard and the carrier, with the battleships, cruisers, and destroyers of the Fleet, headed for the target.

The Fleet was spotted and bombed by the Italians well ahead of the attack, yet the Italian Fleet remained at anchor, taking few extra precautions to get ready for the approaching British—an unforgiveable error in judgment which was to prove costly to Italy.

Although reliable, docile, and highly maneuverable, for which reasons it would remain in service throughout the war, the elderly *Swordfish*—the "Stringbag," as it was affectionately known—was a biplane of relatively low performance. As a consequence, a day strike against the Italians was considered potentially too costly to succeed against so well-defended a major target, particularly without the advantage of surprise. Consequently, a night attack was planned. Accordingly, during the hours of darkness of 11–12 November 1940, the first strike of 12 *Swordfish* went off, six with torpedoes, six with flares and 250-pound S.A.P. bombs. In the face of heavy anti-aircraft fire and a balloon barrage, the torpedo-carriers swept in low, getting in two telling strikes on the battleship *Littorio,* and a third which failed to explode. Another torpedo struck the *Conte di Cavour,* and she sank in shallow waters. One *Swordfish* failed to return.

An hour later, a second strike of five torpedo-carriers and three flare/bomb droppers went in and scored another torpedo hit on the *Littorio* while others hit the *Caio Duilio,* which had to be beached in a sinking condition. Meanwhile, the bombers of the two waves had hit the seaplane base with great accuracy, wrecking it, setting fire to oil storage tanks, and hitting other warships. One more *Swordfish* was lost.

For the loss of two aircraft, the crew of one of which survived to become prisoners of war, the Royal Navy had put half the Italian battle fleet out of action. *Conte di Cavour* was subsequently raised, but never returned to service. *Littorio* and *Caio Duilio* were out of service for over six months. The rest of the Italian Fleet was promptly withdrawn to northern bases, and seldom put to sea again when British aircraft carriers were known to be operating. It was a staggering victory—both material-wise and morale-wise—at a time when the British needed just such a success. It was also the first successful major strike by carrier-based aircraft on an opposing fleet. It is believed by many that the results of Operation Judgement formed the blueprint for the Japanese to refer to when developing their attack plans on the American Fleet at Pearl Harbor.

About two months after the stunning success over the Italian Fleet, *Illustrious* limped across the Atlantic for major repairs in a U.S. shipyard, smashed almost to destruction by German dive-bombers. Such are the fortunes of war!

Supermarine *Spitfire*

It was the general intention of the R.A.F. during the Battle of Britain that the more numerous Hawker *Hurricanes* should concentrate their attention on the German bomber stream while the high-performing Supermarine *Spitfire* should engage the escorting Messerschmitt Bf 109E fighters at high altitude. While in practice this was not always possible, the plan nevertheless worked well most of the time, and the remarkably evenly matched *Spitfires* and Messerschmitts fought many savage duels in the sub-stratosphere over southern England.

By November of 1940, the daylight phase of the Battle was basically at an end, with the bombers now braving the resilient British defense only at night. Nonetheless, the German *Jagdflieger* (fighter pilots) still continued to come over at great height during the daylight hours, either escorting a few "nuisance" fighter-bombers, or on *Freiejagd* (free chase) sweeps that were designed specifically to entice the British fighters up to give battle, and to inflict continuing losses upon them.

Most of the German pilots had been in action almost constantly ever since the previous May 10th, and their scores were mounting. Three men were ahead of all the rest: Majors Werner Mölders, Adolf Galland, and Helmuth Wick, each of whom had been promoted to command a full *Jagdgeschwader* of some 70–100 fighters. By late November of 1940, all three had attained their 50th victories.

After trailing the leaders, Wick pushed into first place when he scored his 55th kill during the morning of November the 28th. Wick reappeared in the British skies that afternoon to seek out additional victims. And, after a further victory, he saw a formation of 12 *Spitfires* below him: This was 609 Squadron, which was one of the most successful Fighter Command units of the Battle.

Six Hundred Nine Squadron had been in the thick of the fighting ever since the evacuation at Dunkirk in late May. This almost constant involvement in the air war resulted in the pilots becoming extremely experienced in combat, especially against the Bf 109 pilots. Leading the British formation was Squadron Leader Michael Robinson, himself an ace, who was to ultimately claim 19½ victories.

Flying one of the other *Spitfires* was Flight Commander Fl. Lt. John Dundas, who, at the time, was the squadron's top scorer with 12½ victories. Dundas had been a pre-war, "part-time" Auxilliary pilot. He was an urbane, sophisticated, and highly educated man as well as an intellectual and gifted sportsman. Dundas was a very popular person.

Upon spotting the unsuspecting *Spitfires,* Wick, a dedicated professional airman, led his *Schwarme* of four Messerschmitts down to attack, but was seen in the nick of time to allow the break to be called by Robinson.

Wick caught a *Spit* in his sights, fired, and saw his rounds slamming into it. Dundas was on Wick's tail at once and began firing, causing mortal damage to the Messerschmitt (which is our scene recording *When Aces Meet*). "Whoopee! I've got a one-oh-nine!" Dundas enthused to the others. "Good show, John!" Robinson shot back, and even as he spoke Dundas's *Spitfire,* and that of his wingman, P.O. Ballion, were catching rounds that sent them plummeting to follow after Wick's machine. The three bailed out of their stricken aircraft near the Isle of Wight, off the southern coast of England, and while blossomed parachutes were seen, neither Dundas nor Wick were ever found; only Ballion survived the combat.

One of the victors in this melee was Lieutenant Rudi Pflanz, Wick's Number 2. Pflanz was already an ace when he participated in the engagement. He was a tough pilot and would have certainly claimed his

52nd victory on the Western Front had he not met his death in combat —with *Spitfires*—in July of 1942.

And so it went!

Gloster *Gladiator*

As a contemporary of the U.S. Navy's Grumman F3F fighter family, the Gloster *Gladiator* was one of the last classic biplane fighters.

This plane was to be immortalized in the story of *Faith, Hope, and Charity,* which were the names of a trio of Gloster *Gladiators* that were supposed to have defended the island of Malta when they rose into the sky to do battle with hordes of attacking Italian bombers. To a great extent this thrilling story was a piece of brilliant wartime propaganda designed for home island consumption, for the fact was the achievements of the *Gladiators* stationed on Malta were very limited. At the time, enemy raids on the island were quite infrequent. Moreover, the defenses were soon reinforced by tidy numbers of the more potent *Hurricane* fighter planes. Despite this, the *Gladiator* did earn its place in aviation's "Hall of Fame."

In October of 1939, at the beginning of World War II, this spiffy little fighter plane was still operational in Britain and France. One can only imagine the gut-wrenching feelings of her pilots once they became knowledgeable of the awesome fighting capabilities of the power-laden Messerschmitt Bf 109's: Chariots pitted against tanks! Well, it had happened before.

Gladiator pilots fought the hard fight over Norway in 1940, operating some of the time from the surface of a frozen lake. It was in the Mediterranean theater, however, where the *Gladiator* was to truly make its mark: not so much in the skies over Malta, as the story went, but over Greece and in the parched sky above the sun-baked desert of North Africa, where the sturdy plane provided the R.A.F.'s sole fighter equipment during the beginning months of the war, holding the line until its bigger, more sophisticated fighter cousins could arrive. *Gladiators* also flew from the heaving decks of Royal Navy aircraft carriers to provide air defense for the Mediterranean Fleet during the summer of 1940, a task the plane performed with great success.

The obsolete fighter was always outclassed by the German machines; her pilots ached to become embroiled in an even match, to test once and for all time the mettle of the machine and pilot . . . and they knew there was probably only one place on the planet where the test could be made—in the sky above the Western Desert of North Africa. If the *Gladiator* was to leave its mark, then it would be here.

On 8 August 1940, 14 *Gladiators* of 80 Squadron, R.A.F., lifted up from the desert sands to form up and fly an early evening sweep over Italian territory.

They were not alone in the desert sky.

A flight of 16 Fiat CR-42 *Falcos,* also biplane fighters, from the 9 and 10 *Gruppo* of the 4 *Stormo C.T.* of the *Regia Aeronautica* were flying cover for Meridionali Ro 37bis reconnaissance biplanes on a mission to the front line.

The two opposing forces met in the area of El Gobi near the frontier between Egypt and Libya, and the well-matched formations clashed violently. The long awaited test was on!

This, then, had to be *the* air battle for us to capture and record on canvas. Because it represents a milestone, perhaps the very last one in the history of the biplane fighter as a combatant, we have recorded the action from two viewpoints and corresponding paintings. The "sister" scene to this one appears later in the book when we cover the *Falco.*

Back to the battle near El Gobi.

When the raging, twisting, and turning fight was over, and the dust and smoke had cleared, seven of the Italian fighters had gone down, against an R.A.F. claim for nine kills and six probables. This against 80 Squadron's own loss of two *Gladiators.* The mark had been left!

Among the successful R.A.F. pilots, several of whom were destined to become future aces, was Flight Lieutenant M.T. Pattle, who would later be considered as the probable top-scorer in the R.A.F. Pattle is believed to have raised his total victories to 50 before his untimely death in Greece in 1941.

The R.A.F. was not the only user of *Gladiators.* Other air forces also employed the plane: South Africa, Norway, Finland, Belgium, and Greece. The fighter was also seen in the China skies during that nation's struggle with the Japanese in the late 1930's. *Gladiators* also served peacefully with the air forces of Latvia, Portugal, Egypt, Eire, and Sweden. This last nation actually sent a volunteer unit armed with *Gladia-*

tors to aid the Finns in their winter war with the Soviet Union late in 1939.

The *Gladiator* was indeed a true international fighter.

Bristol *Beaufighter*

The *Beaufighter* was developed from the *Beaufort* torpedo-bomber. It was a big, heavy aircraft—one of the largest and heaviest twin-engined fighters to see service during World War II.

The plane was not particularly fast or maneuverable, but it was sturdy, reliable and, best of all, packed a terrific punch, with four 20mm cannons in the nose and six .303 machine guns in the wings, mounted (oddly) four in one wing and two in the other! The *Beau's* very size provided one of its secrets for being able to wreak massive damage on just about any target trapped in the gun-sight. The radar operator/navigator, who was seated well back in the capacious fuselage, could reload the cannons while in flight—a unique advantage for a WW II fighter plane!

Because of its size, the *Beaufighter* was able to carry the first airborne radar sets, which were both bulky and unwieldy. Therefore, the plane's first combat role upon entering service was that of a night fighter and, as such, the *Beau* became the most effective element in England's defense during the terrible *blitz* in the winter of 1940–1941. Then, as production increased, radarless *Beaufighters* began operating as long-range day fighters. The *Beau's* first job in this capacity was with the R.A.F.'s Coastal Command, with whom the assignment was to hunt and destroy German anti-shipping bombers snooping for targets around the coasts of Britain and in the Bay of Biscay. The *Beau* was, however, soon sent to the Mediterranean, where she was to first operate out of Malta with the assignment to use her guns to disrupt the Axis communications network, which stretched from southern Europe to the North African coast.

Following the Anglo-American landings in French North Africa in 1942, the *Beaufighters* of 272 Squadron, flying from the island of Malta, played great havoc with the stream of Axis transport aircraft on ferry missions across the Mediterranean from Sicily to Tripoli or Tunis. One of the highlights of this period—and it is the subject of our painting—occurred on 24 November 1942, when the Australian *Beaufighter* ace, Flying Officer "Ern" Coate, shot down a giant six-engined Blohm and Voss Bv 222 flying boat. His victim was one of the prototypes of this aircraft that had been pressed into service for the emergency, and was Coate's fifth victory of a final total of 9½.

Then, operating from airfields in the desert, *Beaus* struck at Axis airfields, convoys, and other targets of opportunity, inflicting great punishment on the enemy wherever he was found. Night fighter configurations quickly joined the day fighters to counter the enemy's night bombing raids on Suez and Malta . . . at every turn the enemy seemed to be met by a *Beaufighter* of one kind or another!

And it was not long after the initial Japanese assault on Burma in early 1942 that *Beaufighters* arrived at bases in India. From here the mighty *Beaus* went out by day to strafe the long and tortuous Japanese lines of communication in Burma and then to provide night defense for Indian cities such as Calcutta.

Under license, Australia built a sizeable quantity of *Beaufighters* and employed them in operations—mainly in the anti-shipping role—around the islands of the South-West Pacific. It was here they flew alongside the aircraft of General Kenney's 5th Air Force in the never-ending quest to relentlessly beat back the enemy.

As the war progressed, *Beaufighters* were adapted to effectively perform other combat roles dictated by new tactics brought about by the ever-changing face of war. This included being first loaded with bombs, then rockets, then torpedoes. Because of her adaptability, the *Beau* remained in action to the very end of the hostilities.

The versatile *Beaufighter* was even supplied under reverse Lend-Lease to several U.S.A.A.F. squadrons operating out of Italy. The Americans were quick to use them for many jobs—as both day and night fighters, against shipping in the Adriatic, and in support of Tito's partisans in Yugoslavia. And the *Beau* continued her anti-shipping role back home by flying with the Coastal Command's Strike Wings around the coasts of North-West Europe and Norway.

Avro *Lancaster*

Until the first atomic bomb was detonated, the dream of many inventors of WW II weapons was to devise a weapon that, with a single blow, would effectively shorten the war.

One such attempt was made by the brilliant innovator and aircraft designer, Dr. Barnes Wallis, after he laid a curious eye toward the German Ruhr. Four mighty dams spanned the Ruhr: the Möhne, Eder, Sorpe, and Enneppe, and each held back tons upon tons of water used to supply hydroelectric power to Ruhr Valley industry. Wallis reasoned that if these dams were broken, the resulting flood would not only be devastating to those caught in the Ruhr at the time, but would also cause a disastrous loss of prime power, vital to the war industry located there. Certainly! Break the dams and the war would be shortened.

Now the question "How to do it?" formed in Wallis's active mind. He knew that for every problem there is at least one answer. He knew conventional aerial bombardment was out of the question: Waves of bomber formations would be required, and the losses to enemy flak and fighters would not be worth the results to be gained. Neither would it do to send in planes carrying super-powerful torpedoes: The crafty Germans had already covered themselves by spreading torpedo nets in front of the dams. There was only one practical way to do it, and that was to skip a powerful bomb across the water, using the same principle small boys employ to skim flat stones across a pond or stream. Bounce the bomb to end its momentum against the dam, next to which it would sink and then explode, and the dam would break! A devastating flood would follow.

First, a weapon was needed. Wallis devised a large cylinder packed with RDX high explosive. The brute he built weighed in at nearly 10,000 pounds!

Then, the weapon carrier. Not much of a choice here. It could be carried only by the four-engined Avro *Lancaster* bomber, the greatest load-carrier of the war. Accordingly, specially-modified bombers were delivered to a newly-formed unit—the 617 Squadron—which was destined to go down in history as The Dambusters!

Now came the resolution of the variables: At what height and distance do we launch the weapon to cause the cylinder to bounce and stop next to the dam? Through trial and error all of this was worked out to perfection, but not before some ingeniously simple devices were created to overcome the inadequacy of flight instruments available at the time.

Again, the bomb. The rascal wouldn't behave the same way each time it was launched. This erratic performance was tamed when it was discovered that it would track a true path, bouncing a constant number of times en route to playing out, when it was released, spinning, at 500 r.p.m.

The next problem was the distance for release. Releasing too early or too late would result in a good try only, for any damage inflicted under either of these conditions would be negligible. The distance problem was solved by mathematics and a very simple optical device. The distance between the two fixed parapets on the dam was known. A simple instrument having two pencil-sized uprights, spaced apart mathematically, was made for use by the bombardier. The instrument was positioned before the bombardier's eyes at a mathematically pre-determined distance. When the run-to-target began, the two uprights would be seen on either side of the parapets, and they would then move toward the parapets as the bomber closed in on the dam. The launch point came when the uprights overlayed the parapets.

Then, the exact height necessary to cause the right number of bounces was calculated. This was to be an extremely low bomb run, and as such it had to have error-free execution. Powerful spotlights were mounted under the nose and tail and then trained downward at predetermined angles that caused the beams to cross each other at the exact distance to get the proper bombing height. The two spotlights would be turned on when the run-to-target commenced and an observer would monitor the two spots on the water's surface. As the pilot lowered the altitude, the two spots would naturally move toward each other, and the correct altitude was attained when they converged. The observer would, of course, be in continuous contact with all members of the crew as he called out the movement of the spots.

After several weeks of extensive and highly secret training, the mission was launched during the night of 16–17 May 1943. Nineteen *Lancasters* rose and formed up in three waves of three groups each and flew out low over the sea to escape detection by enemy radar. The first force of nine bombers was led by the commanding officer, Wing Commander

Guy Gibson, a veteran bomber pilot, whose target assignment was the biggest dam—the Möhne. Gibson would bomb first when they arrived at the target. The big bomber came roaring over the lake a scant 60 feet above its calm surface; Gibson's speed, 220 m.p.h. The altitude lights popped on, and upon seeing them the Germans on the dam opened up at point-blank range laying down concentrated light automatic flak. Gibson pressed on, the observer feeding altitude inputs, the bombardier's eyes frozen to his sights. It was a perfect run-to-target with the beams and sights converging when and as they should. The bombardier depressed the bomb release button and the spinning cylinder-bomb fell away to strike the water and begin bouncing (our scene) to mate perfectly with and explode against the dam.

But the Möhne held, and failed to break!

The next on-rushing bomber was hit by flak, breaking the precision required for a perfect launch, and the bomb bounced to hit a parapet and explode without causing any serious damage. (We can only imagine the emotional state of mind present in the Germans manning those light automatics!)

The third attacking *Lancaster* dropped short, but the next two in were spot on target and with a mighty roar the concrete and masonry of the dam crumbled, sending a great jet of pent-up water thundering into the valley below.

The Möhne had been broken!

But not without the British paying a price: One of Gibson's *Lancs* was lost to flak on the way in. Now Gibson led the remaining still-armed *Lancasters* to the Eder Dam. The first *Lancaster's* run-to-target was too low and it struck a parapet and was destroyed from the blast of its own bomb. The next two, however, made letter-perfect releases and the Eder, too, was split asunder.

A host of problems consumed the other two formations, each of which were composed of five *Lancasters*. One could not find its target and returned home with the bomb intact; two had returned early after suffering heavy flak damage; four others went down. The three remaining *Lancasters* attacked the Sorpe Dam and failed to break it, then the Enneppe Dam without effect. One more *Lancaster*, flown by deputy commander Sqn. Ldr. H.M. Young, was lost on the return flight over Holland. In all, seven of the 19 *Lancasters* failed to return.

The damage was extensive. Ruhr coal mines, power stations, and factories up to 40 miles away were flooded. Communications were disrupted and hundreds of people were drowned, but the results to the German war industry were far less devastating than had been hoped for, and the dams were subsequently repaired. A repeat attack could not be made, for the necessary defensive measures involving steel nets, increased flak, balloon barges, and smoke screens were quickly and efficiently instituted by the Germans.

Wing Commander Gibson, who received the Victoria Cross for his leadership of this daring attack, is seen in our painting, which captures the moment after the bomb was launched. Months later, after completing several tours both as a bomber and night fighter pilot, Gibson was assigned to a staff post. He managed to beg for "just one more raid, this time in a light *Mosquito*"; Gibson did not return.

Handley-Page *Halifax*

While overshadowed by its famous compatriot, the *Lancaster*, the *Halifax* preceded the latter's entry into service with the R.A.F.'s Bomber Command by several months.

Halifaxes were the second British four-engined bomber to be ordered in quantity, and they made their first operational sorties during 1941. The *Halifax* rapidly became the second most important British bomber in terms of numbers, for no less than 6,176 of these big aircraft were to be built.

The bomber was steadily developed and improved, and the later versions to enter service bore little resemblance to the early production models. Powered variously with Merlin in-line engines or Bristol Hercules radials, the *Halifax,* in its several models, flew in excess of 82,000 sorties with Bomber Command, and during these sorties suffered 1,833 operational losses. Her bombardiers did, however, drop nearly a quarter of a million Imperial tons of bombs on the enemy.

Thirty-four Squadrons of Bomber Command were to use the *Halifax* during the war over Europe. Two of these squadrons were units of the French *Armee de l'Air* which had been attached to the R.A.F. By the latter part of the war the bomber was being used mainly by No. 4

Group and the all-Canadian No. 6 Group, with the former flying large number of *Halifax* Mark B.VI's and the latter Mark B.VII's. During June of 1944, the *Halifax*-equipped No. 4 Group achieved the Command's best success rate of the war when the Group's gunners claimed 33 intercepting night fighters shot down that month.

Other *Halifaxes* served in the Mediterranean from 1942 to 1944. Here they raided targets in Libya, Tunisia, Crete, Sicily, Italy, and the Balkans. More served with Coastal Command, functioning as long range anti-submarine patrol bombers which went out to sweep the vast wastes of the Atlantic.

The great majority of the *Halifax's* work was undertaken at night during 1943–44, but by the end of the latter year the virtual disappearance of the Luftwaffe from in the skies over Western Europe allowed the bombers to begin making large-scale daylight attacks, similar to those being flown by the U.S. 8th Air Force. When this change from night to day occurred, wings of R.A.F. *Mustang* fighters were formed to give escort and provide protection to the relatively lightly armed British "heavies."

Accordingly, interceptions by the Germans were indeed rare. Exceptions did occur, however, when on odd occasions some of the formidable new Messerschmitt Me 262 jet fighters would appear out of nowhere and manage to break through the fighter escort to rip into the bomber formation. It is this duelling between *Halifaxes* and Me 262's that we selected to record with a painting.

Hawker *Tempest*

In 1940 the Hawker *Hurricane* was one of the heroes of the Battle of Britain, but by then it was already past its prime. Sidney Camm, the designer of this famous fighter, set out to produce a new high performance interceptor designed around a massive new 2,000 h.p. engine. The result was the *Typhoon,* a powerful monster, dragged along at over 400 m.p.h. by the 24-cylinder Napier Sabre engine. Yet, while the *Typhoon* would later become one of the finest fighter-bombers of the war, as an interceptor the aircraft quickly proved to be a flop!

Wind tunnel tests soon showed the reason for the *Typhoon's* lack of performance at high altitude—Camm's wing design had closely followed that of the *Hurricane* and was far too thick in section, producing the unacceptable levels of drag above certain altitudes. Back on the drawing board, Camm refined his design with an elegant, ultra-thin wing of elliptical planform, which he married to a cleaned-up fuselage and larger tail surface. The result was the *Tempest,* a superlative fighting machine that was put into production with all possible speed.

Featuring high speed, good maneuverability, a heavy armament of four 20mm cannons buried in its graceful wings, together with massive power for the climb and adequate weight for the dive, the *Tempest* was unequalled as a low and medium altitude fighter during the last year of the war in Europe.

This new fighter entered service in the spring of 1944 and had its first taste of combat during the Normandy invasion. Within a matter of days, however, all available *Tempest* units were pulled back to combat the deadly V-1 "doodle-bugs" that were then appearing over southern England. The *Tempest* proved to be one of the few fighters able to catch the bombs in level flight, and thus became their main executioners. Squadron Leader Joseph Berry, top scorer against the bombs, with 60 of them shot down, gained all of his successes while flying the *Tempest,* as did many of the other leading "bomb-busters."

At the end of September 1944, with the worst of the threat from the V-1's past, the *Tempest* was released to again operate over the front lines, and five squadrons joined the 2nd Tactical Air Force in Holland and Belgium. The *Tempests* were very active in the fighting over the Ardennes during the famous Battle of the Bulge in late 1944, but had their heyday during the spring of 1945. In a useful partnership with the high-altitude, Griffon-engined *Spitfire* XIV's, they hunted the dwindling Luftwaffe throughout northern Germany, exacting a heavy toll each time the enemy came up to fight. Throughout the final seven months of the war, the *Tempests* were employed purely as air superiority fighters, patrolling in the areas where they were most likely to bring the Luftwaffe up to do battle, and haunting the airfields from which the Luftwaffe's jet-powered Messerschmitt Me 262's were operating. In the process, they claimed the honors as being the R.A.F.'s best killer of these formidable new opponents. When the war ended in May of 1945, *Tempest* strength in Europe had been increased to seven squadrons.

One of the greatest exponents of the fighter was New Zealander Evan

Mackie, who ended the war leading the largest *Tempest* Wing. Mackie gained the last 5½ of his 21½ victories flying these aircraft, and also used it in destroying other German planes on the ground. But his combat on 7 March 1945 was remembered by him as his hardest of the war. In a twisting, turning fight with a "long-nose" Focke-Wulf Fw 190D, Mackie was literally bathed in sweat before he managed to finally put a burst into the enemy fighter at maximum deflection, sending it blazing down to crash and explode. It was Mackie's 19th victory, and it is this epic combat that is recorded in our scene. At the time, Mackie was commanding 80 Squadron—one of the R.A.F.'s top-scoring units of the war.

Hawker *Typhoon*

"Never enough time to do it right, but always enough time to do it over!" This age-old adage was surely uttered on more than one occasion in regard to the *Typhoon,* for here was an aircraft, designed as an interceptor for use by the R.A.F., that was nothing but a heap of trouble when it first entered service in 1942.

The root of the *Typhoon's* problems formed when it was rushed into production before the airframe had been fully proved and the new and tempermental 2,000 h.p. Napier Sabre 24-cylinder in-line engine had been fully readied for service. As the result, the plane showed a disappointing performance at higher altitudes where the official design called for it to operate. Moreover, the premature introduction into service caused a spate of nerve-shattering accidents to occur: Either the tail would break off, or the engine caught on fire, seized up, or leaked poisonous fumes into the cockpit! For a time it looked as though the *Typhoon* would have to be withdrawn.

It was the Germans and not necessarily the British who gave a second chance to the *Typhoon.* The reprieve came in early 1943 when Focke-Wulf Fw 190 fighter-bombers began a series of low-level hit-and-run raids on the southern coast of England. The *Typhoon* was pressed into service to interrupt this menace and quickly proved it was the one aircraft in the R.A.F.'s inventory that could catch the raiders at this height. It was now clear that the *Typhoon* was no good as a high altitude fighter, but at low levels it more than held its own with the best the Luftwaffe could pit against it.

Accordingly, Hawker strengthened the tail surface, then attacked the plane's many other maladies to correct them gradually, but the Sabre engine always had to be treated with care.

Then, in 1943, an entirely new and promising future opened up for the *Typhoon,* when bombs were fitted beneath the wings to try her as a fighter-bomber. The *Typhoon* reached the pinnacle of success in this classic combination when, later in the year, launching rails to accommodate eight of the new 3-inch rocket projectiles were fitted beneath the wings. This muscular arsenal, coupled with the existing armament of four 20mm cannons, gave the *Typhoon* enormous hitting power.

In the months prior to the invasion of Normandy, *Typhoon* fighter-bombers swept the coasts of France and Belgium, bombing and rocketing and strafing radar stations, knocking out bridges, mangling rail targets, airfields, and coastal defenses, shooting up convoys and other like targets, completely terrorizing the enemy at every turn.

When the invasion was launched in 1944, *Typhoons* were among the first aircraft to be moved onto airstrips within the beachhead. Eighteen squadrons of these deadly aircraft provided the R.A.F.'s main, close support strength. And they were active from the onset, quickly establishing their position as oppressors of the German Panzer Divisions.

Throughout the remainder of the war, the *Typhoons* brought their awesome firepower to bear on the Germans: Breaking up armored attacks, winkling out 88mm guns, *Nebelwerfer* multi-barrel mortars, self-propelled guns, Tiger tanks, and other strong points holding up the Allied infantry or tanks during the offensives, playing havoc with the German columns at Falaise, and generally creating mayhem. When the front was quiet, they ranged behind the German lines on interdiction missions, hunting transport columns, rail locomotives, barges, and other river traffic which were attacked with cannon, rockets, and bombs.

Although not employed on any other front, the *Typhoon* was without any doubt *the* British fighter-bomber of the war.

Gloster *Meteor*

The Gloster *Meteor I* was the first Allied jet-powered aircraft to see combat duty during World War II.

It was less of a great leap forward in aeronautical progress than was the first production jet, the Messerschmitt Me 262. The *Meteor*'s performance and armament was similar to that of late war, piston-engined fighters. Nonetheless, the *Meteor* pointed the way to the many advances in the state-of-the-art made possible by the jet engine.

*Meteor*s were pressed into service with the R.A.F.'s 616 Squadron during the summer of 1944, when the Anglo-American forces were struggling to establish their beachheads in Normandy. Initially, the new jet fighter was employed only to intercept and destroy the V-1 flying bombs that were raining down on southern England at the time. The *Meteor* proved to be one of the few aircraft capable of catching the "buzz bombs," or "doodle-bugs," as the British were prone to call them, in level flight. Normally, the *Meteor* pilot would select his target, give chase to close in, and shoot it down. Occasionally, a pilot would, for one reason or another, fly right up to the doodle-bug and maneuver a wing-tip to below the bomb's stubby horizontal wing, then flip it (the subject of our painting). This threw the bomb's guidance system out of control and the bloody thing would then crash and explode harmlessly.

It was not until late in 1944 that a few of these aircraft were moved to join R.A.F. units stationed in Holland. Their combat assignment in this rear area was to provide defense against Luftwaffe attacks, should any occur. None ever materialized.

Then, in the spring of 1945, when the war was in its final agonies, *Meteor*s were released for operations over the front lines. By this time, of course, the once mighty Luftwaffe had been decimated, and it was for this reason that the *Meteor* pilots were to never engage German aircraft in combat in the sky. They did, however, manage to strafe a few on the ground.

Shortly before V-E Day, further squadrons of improved *Meteor Mark III*'s, along with other squadrons flying the new DeHavilland *Vampire* jets, were ready for action, and they would have undoubtedly played a major role in the air war had the hostilities lasted longer.

The *Meteor* was, however, destined to see aerial combat. The chance came during the conflict in Korea, where Australian pilots flying the much-developed *Meteor Mark VIII*'s did battle with the Russian MIG's. Other *Meteor*s also saw limited action in the Middle East, where they were operated by both Egypt and Israel.

THE MIXED GALLERY

Now come twelve electrifying scenes, each depicting the markings of a mixture of six nationalities locked in mortal combat—Russia, Poland, France, Italy, the Philippines, and the United States.

It is on these following pages where you will get a glimpse of combat aircraft that is rarely seen in books, and never before seen in battle. It becomes a thrilling experience to move back in time and sit behind these pilots, to look over their shoulders and see what they saw . . . you can almost hear the engines growling, feel the plane vibrating beneath you, shuddering with the blasting of the guns, the winds of war whistling past you when your gull-winged P.Z.L. P-11 dives into a formation of Dorniers, as you scream down on the deck in a *King* to rip into German armor, twisting and turning in *Ding Hao!* as you defend *Fortress*es under attack by Bf 110's and 109's. Ah yes . . . it sets the adrenalin to pumping, this fighting in the sky. Would you dare pit the guns of your obsolete little *Peashooter* against those of the deadly *Zero*es and *Nell*s? Lt. Jesus Villamor and his small band of courageous Filipino pilots did . . . and scored!

The mighty air arms of the Germans, British, and Americans tend to overshadow those of other countries, whose pilots and planes made their contributions, brief-lived as they may have been. Their duels in the sky have a way of getting lost and forgotten among the countless number of more famous air battles. The *Mustang*s and Bf 109's and *Spitfire*s have claimed center stage, shoving aside these no less gallant warriors, who rose to fight and have their moment of glory.

Now let's read about and see twelve such moments of glory, the original canvases of which are part of the Douglas Champlin private collection on display at the Falcon Field Fighter Air Museum in Mesa, Arizona.

Northrop P-61 *Black Widow*

The Northrop P-61 *Black Widow* was a latecomer to the ETO. The big, twin-engined night fighter still had difficulties to overcome following its arrival in England. The night fighter crews of the 422nd Night Fighter Squadron of the 9th Air Force had to almost give up their planes and, instead, used British-built DeHavilland *Mosquito*es.

Lt. Colonel O.B. Johnson, commander of the 422nd Squadron, had other ideas for his unit. The *Black Widow*s were stripped of the big power turret atop the fuselage and all other excess weight. In this "stripped down" configuration, the aircraft was pitted against the saucy R.A.F. *Mosquito*. In what had been an anxiously awaited competition between the two machines, the big P-61 really showed her stuff. The *Black Widow* proved to be faster than the *Mosquito* at altitudes from 5,000 to 20,000 feet, easily out-turned the British craft, and left it wanting in rate-of-climb tests. The men of the 422nd were ready for a taste of combat.

The *Black Widow*s began their scoring after their arrival in France in August of 1944. The determination and perserverance of the crews of the 422nd knew no bounds, for once directed to a German intruder by the ground controller, the radar observer tracked down his prey on his primitive radar scope, steadily closing on the target until the pilot could bring the aircraft in position to shoot it from the sky.

While flying a defensive patrol over Belgium during the night of 4 October 1944, radar observer Lt. Robert A. Tierney picked up a contact three miles out and at an altitude of 21,500 feet. Under Tierney's guidance, the pilot of the *Black Widow*, Lt. Paul A. Smith, followed the enemy plane through its mild evasive action up to an altitude of 24,000 feet, where visual contact was made.

Smith pulled "Lady Gen" up to within 400 feet directly astern and slightly below the positively identified Messerschmitt 410. After one short burst and one long burst, strikes were observed on the fuselage and wings, followed by a small flash, then a large explosion as the target dropped off on his back, debris falling off and thick smoke pouring from both engines. The Messerschmitt was then observed in a vertical dive which culminated in a violent explosion when it crashed into the ground.

Smith and Tierney went on to become the first of three "ace" crews of the 422nd Night Fighter Squadron. By the time the war ended in Europe, the P-61's of the unit had destroyed 43 enemy aircraft in the air to become the most successful A.A.F. night fighter squadron of World War II. The *Black Widow*s of the 422nd scored victories ranging from airspeeds of 375 indicated at 24,000 feet (against the Messerschmitt downed by Smith and Tierney) to a Junkers Ju 52 which was shot down at 90 IAS at 1,000 feet. In addition to enemy aircraft, the unit was also credited with five V-1 flying bombs, numerous locomotives, rolling stock, and ground installations.

The lethal Northrop fighter left a little-known but nonetheless indelible mark in the history of night fighter operations during World War II.

Brewster F2A-3 *Buffalo*

The Japanese forces were riding high in the spring of 1942. They had taken Malaya, the Philippines, Wake Island, Guam, and had suffered what they considered to be only a temporary reversal at the Battle of the Coral Sea.

Japanese Admiral Isoroku Yamamoto knew that he must make a move to draw the American fleet out into a battle so it could be destroyed once and for all. To accomplish this, he sent a feigning task force toward the Aleutian Islands off Alaska, while his main force steamed toward the American base at Midway Island.

Three lethal Japanese forces set out for the tiny island: An occupation force of troop transports, escorted by a light carrier, battleships, cruisers, and destroyers; a main striking force, built around four large carriers; and the third force, which consisted of battleships, cruisers, and destroyers.

The Americans had one very important initial advantage: The breaking of the Japanese cipher code enabled them to learn of the impending attack, and thus divert all of their meager forces toward Midway.

While the United States fleet sped to meet the rapidly approaching enemy force, the pitiful garrison on Midway girded themselves for the coming battle. The primary island defense lay in the hands of the one and only fighter squadron based on that scant piece of real estate: United States Marine Corps VMF-221, under the command of Major Floyd B. Parks. The unit was equipped primarily with 20 outdated Brewster F2A-3 *Buffalo*es. The original model of the aircraft had performed well; it was maneuverable, carried four .50-caliber machine guns, and was a stable firing platform. Subsequent modifications had, however, increased its weight to the point that its maneuverability had become poor and its rate of climb pathetic.

VMF-221's brief moment of glory came on the morning of 4 June 1942, when a U.S. Navy patrol bomber sighted the first task force of the Japanese fleet approaching Midway. Major Parks led a force of seven *Buffalo*es and five Grumman F4F *Wildcat*s out to intercept a formation of enemy aircraft that had been launched from the task force and was reported "heading directly for Midway." A second flight of seven *Buffalo*es and one *Wildcat,* under the leadership of Captain Kirk Armistead, were vectored ten miles out to await possible action coming from another direction.

The fighter pilots with Parks were 30 miles out and at an altitude of 14,000 feet when they sighted many enemy *Val* dive bombers 2,000 feet below them and, below the bombers, their screening fighters. The Japanese were so confident of finding Midway asleep that they planned to use the *Zero* fighters only for strafing; hence, the lower altitude.

The *Buffalo*es peeled off and went down after the dive bombers. Several of the enemy craft fell to the Brewsters, but their moment of triumph was short-lived. Once they passed through the tight vee formation of bombers, they were set upon instantly by the escorting *Zero*es. One after the other the *Buffalo*es were either shot down or cut to ribbons by the Japanese fighters. Even when the fight was joined by Captain Armistead's forces, the Marines still suffered their worst defeat of World War II.

The Japanese lost a few of their aircraft to the *Buffalo* pilots during this fight, but when the combat was over and the Americans counted their losses, 13 of the Brewster *Buffalo*es and two of the *Wildcat*s were

missing. Of the few that were able to limp back to Midway, only two were airworthy, and neither of these were Brewsters.

The Battle of Midway was a resounding triumph for the U.S. Navy and proved to be the stroke that broke the back of the Japanese fleet, but it was the darkest hour that the men of the Marine Corps fighters had ever known. The qualities their aircraft lacked for their initial combat was made up for by the sheer guts of the pilots of VMF-221. Speaking of the loss of his leader, Major Parks, and all the others on that day, Captain Philip R. White stated: "It is my belief that any commander who orders pilots out for combat in an F2A should consider the pilot as lost before leaving the ground."

Boeing P-26A *Peashooter*

Boeing's P-26A was the last of the U.S. Army Air Corps *Peashooter*s. Its predecessors during the late 1920's and early 1930's had all been biplanes, so the arrival of the P-26 on the scene in 1934 made it the pride of the service.

The stub-winged little craft was the first all-metal monoplane fighter accepted by the Army Air Corps. Initial assignment was made to the 20th Pursuit Group at Barksdale Field, Louisiana, the 17th Pursuit Group at March Field, California, and to the 1st Pursuit Group at Selfridge Field, Michigan. The performance and maneuverability of the P-26's made these units the envy of all the pursuit pilots in the Air Corps. The 500 horsepower Pratt and Whitney R-1340 engine propelled the craft along at 211 miles per hour top speed at sea level and achieved a maximum speed of 324 miles per hour at an altitude of 6,000 feet. It was armed with two .30-caliber machine guns that were mounted internally in the fuselage and fired through the propeller.

When war broke out in Europe in 1939, the colorful P-26A's were in the process of being relegated to units outside the United States. A number of the aircraft were assigned to the Panama Canal Zone, to Wheeler Field at Hawaii, and 35 were sent to units in the Philippine Islands.

When war clouds began to gather in the Pacific, last minute attempts were made to build up fighter forces in the Philippines. As the newly designated Army Air Force units began to receive Curtiss P-40's, some of the now very obsolete little P-26's began to trickle into the 6th Pursuit Squadron of the Philippine Army Air Corps. This unit was equipped with nine of the aircraft when the Japanese attacked the Philippines in December of 1941.

The plane that had been the pride of them all in the mid-1930's was in no way capable of taking on the modern aircraft of the Japanese forces, but for the valiant pursuit pilots of the Philippine Army Air Corps there was nothing else available. When Zablan Field came under attack by enemy aircraft on 10 December 1941, Captain Jesus Villamor led a little force of P-26's from Batangas Field to intercept them. While the scrappy Filipino fighters did not down any of the enemy aircraft, they did interrupt the attack and prevented further damage to Zablan.

To have their forces intercepted by such an antiquated enemy was more than the Japanese could stand. On 12 December 1941, 27 Mitsubishi *Nell* bombers, escorted by *Zero* fighters, mounted an attack on Batangas Field. However, when the enemy bombers arrived, a force of five P-26's under the able leadership of Captain Villamor was already airborne and waiting for them.

The *Peashooter*s were able to make one diving pass at the Japanese bombers before they were set upon by the escorting *Zero*es. Villamor sent one of the *Nell*s spiraling down before he had to begin the fight of his life against the enemy fighters. Then Lt. Antonio Mondigo's P-26 came under fire and the damage inflicted was so heavy that he had to bail out. When the chute blossomed, Mondigo was set upon by enemy fighters. Lt. Godofredo M. Juliano raced over to rescue his close friend by taunting the *Zero* pilots to come after his *Peashooter*. About this time, Lt. Cesar Basa, who was returning from a recon flight, was swept into the action and shot down.

The 6th Pursuit Squadron paid a heavy price for the one *Nell* that Villamor shot down. But their scrappiness and willingness to fight scattered the Japanese bomber formation, and the tough Filipino pilots gained at least another moral victory for themselves, their people, and the free world.

This action between the *Peashooter*s and *Nell* bombers and *Zero* fighters would be the last concerted effort of the Boeing P-26 during World War II, and it was something of a miracle that the last *Peashooter* should be able to leave a mark of victory in a conflict that was so far removed from its original design. The little plane actually claimed a *Zero* a few days later.

Bell P-63 *Kingcobra*

Although the Bell P-63 *Kingcobra* showed a marked resemblance to the Bell P-39 *Airacobra*, the two aircraft were, in fact, two entirely different designs. The *Kingcobra* utilized a laminar flow wing, was a larger aircraft overall, and incorporated the 1,325 horsepower Allison V-1710-93 engine. Armament was composed of a 37mm cannon firing through the propeller hub, two fuselage-mounted, .50-caliber machine guns, and one .50-caliber machine gun mounted in a pod under each wing. Outboard racks under each wing were provided for carrying a 500-pound bomb, and either a fuel drop-tank or another 500-pound bomb could be carried beneath the fuselage.

The heavily armed *Kingcobra* was not lacking in aerial performance, either. The production model—the P-63-A—had a top speed of 410 miles per hour at 25,000 feet and the ability to climb to that altitude in 7.3 minutes. Later models powered with the Allison V-1710-117 engine had a war emergency rating of 1,500 horsepower at sea level and 1,800 horsepower utilizing water injection.

Unfortunately, the *Kingcobra*'s entrance into service with the U.S. Army Air Force came at a bad time. When production models became available in October of 1943, the P-47, P-38, and P-51 pretty well had all the fighter plane assignments under control.

Of the 3,308 *Kingcobra*s built, 2,421 were destined for delivery to the Soviet Union under the Lend-Lease program. While in service with the Russians, the *Kingcobra* won great acclaim as a ground support aircraft. Its heavy armament and ability to absorb punishment and still come home made the plane a great favorite among the Russian pilots.

The *Kingcobra*s arrived in the Soviet Union just in time to participate in the spring and summer offenses by the Russians in 1944. During this period the *King*s played a vital role in helping to contain, if not drive back, German ground forces in the Ukraine. The Russian-flown *King*s would swoop down on the enemy armor, the pilots holding down the gunners by strafing with cannon and machine guns blazing as they selected juicy targets to go after with the bombs. Then, circling back low on the deck, the pilots would continue to add confusion and destruction to the mayhem by attacking until their ammo was exhausted. Relentless assaults such as this left significant gaps in the German defenses, and the Russian ground forces were quick to advance and exploit the situation to its fullest measure.

We have depicted the *Kingcobra* in the type of operation that made it famous: down on the deck, striking with all its fury at German armor. Untold numbers of German tanks met their fate on the wide expanses of Russia when the Russian pilots came roaring in on them to do battle using the formidable Bell attacker.

Some 300 *Kingcobra*s were sent to the French forces in North Africa, but none of them saw combat duty during World War II (actually, they were not uncrated and assembled until after the war was over!). The U.S. Army Air Force did take delivery of 332 P-63's which were utilized in fighter transition programs and in aerial gunnery training.

P.Z.L. P-11

When World War II broke out in September of 1939, and the mighty Luftwaffe appeared in the skies over Poland, the first line fighter force of the Polish Air Force consisted of 128 P.Z.L. P-11 aircraft.

The little gull-winged fighter was originally a design going back to 1931 when the first prototype flew. Delivery of the P-11a to the Polish Air Force began in 1934, and a further development, the P-11b, was built for export to Roumania. The P-11c, which was to see service during World War II, began to enter service in 1935. This version of the aircraft was powered by a P.Z.L.-built Bristol Mercury nine-cylinder radial engine which developed 645 horsepower. The sleek-lined monoplane had a maximum speed of 242 miles per hour at 18,000 feet. It was armed with two 7.7mm machine guns mounted in the fuselage. Some later models incorporated two additional 7.7mm guns mounted in the wings.

The P.Z.L. was an excellent machine in the mid-1930's, but against the Luftwaffe of 1939 it was hopelessly outclassed. Its speed and rate of climb were not sufficient to combat even the German bombers, let alone the power-packed Messerschmitt fighters.

What the Poles lacked in equipment they made up for in intestinal fortitude, for when the bomber formations made their appearance over Poland on 1 September 1939, the little P.Z.L. P-11 fighters were in the thick of the fighting. *The first air victory of World War II was scored by Lt. W. Gnys of the 2nd Polish Air Regiment when he shot down a Junkers Ju 87 dive bomber at 0520 hours on the first day of the war.*

The man who was to become the leading Polish fighter ace of World War II did his initial scoring the following day. This was Stanislaw Skalski of the No. 142 "Wild Ducks" Squadron, who shot down two Dornier Do-17 bombers as his initial victories.

The P.Z.L. fighters fought until practically all of them had been destroyed in combat. Of the 126 confirmed victories scored by the Polish fighter force, practically all were claimed by P.Z.L. pilots. A number of these pilots managed to fly their aircraft to Roumania at the end of the Polish campaign. From Roumania, many of them made their way to France and England, where they continued their battle against the German Luftwaffe.

Douglas TBD *Devastator*

Preliminary action to the Battle of the Coral Sea on 8 May 1942 included some surprises, miscalculations, and over-expenditures of ammunition.

Prior to the big battle, the Japanese pulled a surprise move at Tulagi, a small island 18 miles off Guadalcanal. Admiral Shima's Tulagi invasion group made an unopposed landing May 3rd on the beaches that the U.S. Marines had to win back three months later. (Of the 500 Japanese troops on the island, the Marines buried 427.)

In support of Shima, Admiral Goto's covering group milled about south of New Georgia while Admiral Marushige's support force was sailing some 60 miles farther west. Meanwhile, Admiral Takagi's big carriers, which would eventually tangle with American flattops, were well north of Bougainville, planning to enter the Coral Sea from the east. The real Japanese thrust, the Port Moresby invasion group, destined for New Guinea, was being made ready at Rabaul.

On the morning of May the 3rd, Admirals Fletcher and Fitch, aboard the U.S.S. *Yorktown* and *Lexington,* respectively, were about 100 miles apart and fueling some 500 miles south of Tulagi. At 7:00 P.M. that evening, Jack Fletcher received the report that Australian-based planes had sighted two transports debarking troops off Tulagi, and that five or six Japanese warships were in the area. The American Admiral headed north immediately, determined to strike Tulagi.

Meanwhile, Goto and Marushige were retiring, feeling the island was now secured. Following their easy conquests of the Netherland East Indies, the Japanese expected no counter-attack at Tulagi.

Fletcher maintained his 27-knot clip northward through the night and at dawn, May 4th, launched 12 TBD *Devastator* torpedo planes and 28 SBD *Dauntless* dive bombers. Six F4F-3 *Wildcat* fighters performed combat air patrol over the carrier.

Yorktown's attackers were made up of the TBD's of VT-5 and the SBD's of VB-5 and VS-5.

The VS scouting squadron arrived first over Tulagi and began attacking at 8:15 A.M. It dropped thirteen 1,000-pound bombs, damaging the destroyer *Kikuzuki* and sinking two small minesweepers. The torpedo planes came in five minutes later, launching eleven torpedoes but only knocking off the sweeper *Tama Maru*. At 8:30, VB-5 bombers dropped fifteen 1,000 pounders, with possible minor damage to two ships. All planes were back safely on the *Yorktown* by 9:30.

An hour later, the second strike combined 27 SBD's, each carrying a half-ton bomb, and eleven TBD's. The *Daunt*s damaged a patrol craft and destroyed two seaplanes. The *Devastator*s went through heavy AA fire, and while all were launched, none of the torpedoes hit targets. One plane was lost upon return.

A third attack, at 2:00 P.M., consisted of 21 SBD's which dropped 21 more half-tonners, sinking only four landing barges. That same afternoon, four *Wildcat*s were sent up to knock out the three Japanese seaplanes still anchored in Tulagi harbor. They also spotted the destroyer *Yuzuki* and strafed her with four runs, killing the captain and many crewmen, but the ship got away.

By 4:30 P.M. the Battle of Tulagi was over. What stood out was the over-estimation of the Japanese force and the damage inflicted. A Fleet minelayer was taken to be a light cruiser, the transport for a seaplane tender, the larger minesweepers for transports, and landing barges for gunboats; only the two destroyers were correctly identified. The fliers believed they had sunk two destroyers, one freighter, and four gunboats, forced a light cruiser ashore, and damaged a third destroyer, a second freighter, and a seaplane tender.

Also, the attack proved that much was left to be desired in the field of targeting. They had expended 22 torpedoes, seventy-six 1,000-pound bombs, and about 83,000 rounds of machine gun bullets. Admiral Nimitz is reported to have said: "The Tulagi operation was certainly disappointing in terms of ammunition expended to results obtained," and he re-emphasized ". . . the necessity for target practice at every opportunity."

Lavochkin La-7

The early days of World War II on the Eastern Front saw the Russian Air Force all but driven from the skies by the German Luftwaffe. The Russians had been caught with inferior aircraft and poorly trained pilots. Once the initial onslaught had been overcome, the Russian designers and manufacturers set in motion a massive production campaign that never wavered.

Initially, the men of the Red Air Force had to depend on small quantities of British- and American-built fighters to hold the line until they had sufficient numbers of fighters of their own design. Their pilots received extensive combat training in tactics, and elite units, such as the Russian Guards Regiment, were formed.

One of the more successful designers was Semyon Lavochkin. His Lagg-3 fighter saw action against the Finns in 1941, but its performance there was not particularly outstanding. In late 1941, he took the Lagg-3 airframe and adapted it to take a Shevtosov fourteen-cylinder radial engine. Its performance with this engine was an immediate success. The plane was not only some 25 m.p.h. faster than the Messerschmitt Bf 109F at low altitude, but its low altitude performance enabled it to turn inside the Messerschmitt at will. It was put into production immediately as the La-5 and promptly won considerable fame in its combat against the Luftwaffe.

The last of the Lavochkin designs to see extensive action was the La-7. This aircraft was a refinement of the La-5, with a more powerful engine and three 20mm cannons for armament, rather than the two carried on the La-5. This aircraft was flown by a number of the top Russian fighter aces including Ivan Kozhedub and Alexander Pokryshin.

Ivan Kozhedub was flying his La-7 on a lone reconnaissance patrol on the morning of 15 February 1945, when his eye caught a glimpse of movement over the snow-covered landscape. Then, when the object passed over a bit of woodland and its light gray color betrayed it to be an enemy Messerschmitt M2 262 jet fighter, Kozhedub dived immediately and closed to within 400 yards before the German sighted him. The jet pilot slammed his throttle forward immediately, but by then it was too late: The 20mm cannon shells from the La-7 hammered into the port wing of the German jet and the engine began streaming smoke and flame. Kozhedub fired again, and the 262 went down, cutting a fiery swath in the woods before exploding in a great sheet of flame.

The Red Air Force and its succession of high performance fighter planes, such as the La-7, had proved itself to be as good as any aircraft to see combat in the days of World War II.

North American P-51-B *Mustang*

The daylight bombing campaign of the U.S. Army Air Force reached a climax in the fall of 1943. Without having fighter escort to go with them to targets deep in Germany, there was no way that the heavy bombers could continue without taking prohibitive losses. But, at long last, help was on the way.

The North American P-51 fighter possessed the sleek lines of a real thoroughbred and showed promise of living up to its namesake—*Mustang*—even though it suffered, performance-wise, at high altitudes. Through the concerted efforts of Major Tommy Hitchcock of the U.S. Army Air Force, and the Royal Air Force, one of the most fortuitous

marriages of World War II was consummated when the P-51 aircraft was mated to the Rolls-Royce Merlin engine. The resulting performance of the aircraft forced the U.S. Army Air Force to sit up and take notice, and the plane was ordered in quantity, with the Packard Motor Company building the Merlin under contract.

The 354th "Pioneer Mustang" Group of the 9th Air Force got into action flying the P-51-B in December of 1943. Although the pilots of the group had not flown the *Mustang* before their arrival in England only a few weeks prior to their first mission, their success was immediate and sensational. Not only were the *Mustangs* able to go all the way to the target with the bombers, the aircraft also proved instantly to be superior to anything the Luftwaffe could put in the air against it. From 20 December 1943, when the 354th scored its first aerial victories, it never looked back. A continuous success was to be theirs throughout their war-time history.

The 354th was assigned the task of escorting the heavy bombers to the twin targets of Halberstadt and Oschersleben, Germany, on 11 January 1944. Leading the group was a veteran fighter ace, Major James H. Howard, who had been bloodied in combat while flying over China with the American Volunteer Group, better known as the "Flying Tigers." Howard had taught his pilots well, but on this occasion they were a bit too eager. When enemy fighters were sighted, the flights began breaking off to attack, and the next thing Major Howard knew, he was alone.

Scanning the skies, Howard sighted a bomber formation that was obviously under heavy attack. Without hesitation, he turned his *Mustang "Ding Hao!"* into the fray. For the next half hour he turned in one of the most fantastic performances recorded in the skies over Europe. Time after time, he broke up enemy fighter attacks which had concentrated on the 401st Bombardment Group. The crews of this unit couldn't begin to sing high enough praises for this fighter pilot upon their return to base. They would confirm six victories for him alone.

The modest Howard would only say that he had a "good day" and claimed two confirmed. However, when all the reports were in, everyone concerned took a different view. Major James H. Howard was awarded the Congressional Medal of Honor.

The P-51's would take the bombers to Berlin many times, and to all the other targets that lay deep in Germany where the Luftwaffe had dominated the skies. By the summer of 1944, the *Mustangs* ruled these skies, and before the end of World War II the heavy bombers were able to penetrate the Third Reich at will.

Macchi C. 202 *Folgore*

Undoubtedly the finest fighter to enter combat with the Italian *Regia Aeronautica* during World War II was the Macchi C. 202. This sleek-lined aircraft was powered with an in-line, liquid cooled Daimler-Benz engine which was built under license in Italy as the Alfa Romeo R.A. 100 R.C. 41.

From its initial flight on 10 August 1940, the *Folgore* (Lightning) proved to be a real thoroughbred. Maximum speed was 375 miles per hour at 18,370 feet and the rate of climb was excellent and its maneuverability was superb. Armament was composed of two 12.7mm machine guns firing through the propeller and two 7.7mm machine guns mounted in the wings.

The Macchi C. 202 was initially assigned to units of No. 1 *Stormo* in the summer of 1941. By November of that year the units were in action in Libya. During 1942 the *Folgores* of No. 1 *Stormo* and No. 4 *Stormo* saw extensive action not only against the Royal Air Force in North Africa but also in the skies over the island of Malta. In the course of these air battles the Macchi C. 202 enjoyed considerable success against the first line fighters of the Royal Air Force. The aircraft could turn easily inside the Curtiss *Tomahawks* and *Kittyhawks* and the Hawker *Hurricanes,* and managed to hold its own against the vaunted *Spitfires.*

An Italian ace, whose star began to rise while flying the Macchi C. 202 during the summer and fall of 1942, was *Capitano* Franco Bordoni-Bisleri. Seven of his twelve victories in North Africa were scored during the brief period between October the 20th to November the 7th. His most outstanding feat was the downing of two Curtiss *Kittyhawks* in a wild dogfight on 1 November 1942.

On this day, twelve *Kittyhawks* of No. 250 Squadron, Royal Air Force, had set out to strafe the Mersa Matruh road when they were jumped by four MC 202's of 18 Gruppo. During a ten minute dogfight,

three of the *Kittyhawks* fell victim to the Italian *Folgore's* guns, two of them from Bordoni's.

The Macchi C. 202 continued to see action not only against the Royal Air Force but also against the U.S. Army Air Force, which entered the fight in North Africa in November of 1942. In these combats the *Folgore* enjoyed advantages of maneuverability and rate of climb against the P-38 *Lightning,* while the Bell P-39 *Airacobra* was no match whatsoever.

In September of 1942, a detachment of Macchi C. 202's was sent to Russia to operate with No. 21 *Gruppo.* While they were few in number, the *Folgores* gave a very good account of themselves against the Russian *Yaks* and Laggs.

The last of the Macchi C. 202's saw action against the massive Allied air attacks against Sicily and, later, following the invasion of Italy. When the Italians surrendered in 1943, a few Macchi C. 202's served with the R.S.I. Co-Belligerent Air Force in the north of Italy.

Dewoitine D. 520

Had the French *Armee de l'Air* been actively involved in the air war against Germany, no doubt the Dewoitine D. 520 would have attained a degree of fame comparable to the leading fighter planes of World War II.

The design was the brainchild of Emile Dewoitine and his chief designer, Robert Castello. The trim little monoplane was powered by the Hispano-Suiza 12Y21 twelve-cylinder in-line engine, and made its maiden flight on 2 October 1938. The initial flight was a bit disappointing because a maximum speed of only 300 miles per hour was reached, but a modification program upgraded its performance greatly.

The second prototype flew on 28 January 1939, and was a vast improvement over its predecessor. An engine change to the 12Y31 raised the top speed up to 341 miles per hour. The airplane's rate of climb was excellent and the maneuverability was exceptional.

A first production order in April of 1939 called for the manufacture of 200 D. 520's, to be delivered between September and December of that year. War clouds gathering over Europe prompted a subsequent order for 600 more of the aircraft.

France was at war when the first production aircraft rolled off the assembly line. The first true production machine was significantly different from the prototypes. It was powered by an H.S. 12Y45 engine which was equipped with a supercharger and an electric propeller. The armament package consisted of a 20mm cannon, mounted between the engine cylinder banks, which fired through the propeller hub, and four 7.5mm machine guns, two of which fired through the propeller and the other two from their mountings in the wings.

When the German *blitzkrieg* into France and the Low Countries began in May of 1940, only *Groupe de Chasse* I/3 was equipped with the D. 520. Before the Franco-German armistice on 25 June 1940, GC III/3, GC III/6, and GC II/7 were equipped with the Dewoitine fighter. The extraordinary performance of the aircraft is substantiated by the fact that the limited number of units equipped with the D. 520 accounted for 114 confirmed victories and 39 probables. This feat becomes all the more exceptional when it is considered that delivery of the aircraft was made under dire combat circumstances and that pilots had practically no time to train in them.

After the fall of France, a number of Dewoitine D. 520's escaped to North Africa where they became a part of the Vichy Air Force, which operated under the control of the German Armistice Control Commission. Under these controls, production of the Dewoitine was resumed in France and quite a number of French units were furnished with the aircraft.

The Dewoitine D. 520 saw further action against the Royal Air Force in the Mediterranean and also in Syria during June—July of 1941. When the Americans invaded North Africa in November of 1942, both the U.S. Navy and the U.S. Army Air Force encountered and engaged in combat with the testy little French fighter.

The Germans dissolved the French Air Force in November of 1942, but the Dewoitine 520 continued to be produced for use by the Luftwaffe until the summer of 1944. These D. 520's saw duty as training aircraft in Germany right up until the end of hostilities in 1945.

One of the top French aces of World War II flew the D. 520 with great success. Adjutant Pierre Le Gloan was a member of a flight of two aircraft from GC III/6 that engaged a formation of Italian Fiat CR 42 fighters during the afternoon of 15 June 1940. Le Gloan shot down three of the Italian fighters and Captaine Assolant claimed another before they were called upon to return to their base, which was under attack by Italian aircraft. Engaging this force, Le Gloan shot down another CR 42, and then proceeded to claim a reconnaissance Fiat BR 20 before landing. His five victories in one combat brought him an immediate promotion.

Le Gloan was to score 22 victories, primarily in the D. 520, before his death in 1943 in a flying accident.

Ilyushin Il-2 *Shturmoviki*

On the morning of 22 June 1941, Hitler's Panzer legions rolled into the Soviet Union. Following the lightning successes of the war in the West that had seen German forces quickly overrun the Netherlands, Belgium, and France, there was immediate speculation that the Russians could not hold out until winter.

Fortunately, it was an early and severe Russian winter that halted the German *blitzkrieg* within sight of the Kremlin spires of Moscow. The war came to a halt until the snow melted, and when it did, the ground turned into a quagmire of mud in the spring of 1942.

Slowly and methodically, the German war machine advanced once more, only to grind to a halt in the southern Ukraine at the city of Stalingrad. Initially, the Soviet Air Force as well as the Russian ground forces had reeled and retreated before the Panzer onslaught, but as new equipment became available the complexion of battle in the sky as well as on the ground began to change.

The most significant challenge to the Germans from the Soviet Air Force came in the form of the most heavily armored, ground attack aircraft that would appear in World War II; the Ilyushin Il-2 *Shturmoviki*. The basic design of the aircraft was extraordinary. The forward fuselage surrounding the engine and the pilot was comprised of an armored shell which ranged in thickness from 5mm to 12mm. Armament consisted of two 7.62 machine guns, two 20mm cannons, and rails for eight 82mm rockets.

Initially the aircraft had been single place, but to further protect the pilot the armored tub was elongated and a gunner armed with a 12.7mm machine gun was added. Progressively, the 20mm cannon in the wings gave way to more powerful 37mm cannon. The awesome machine became known to the German Army as *Schwarz Tod* or "Black Death."

The *Shturmoviki* met with immediate success against the German ground columns and even more so as a tank destroyer. Its cannon and rockets took a tremendous toll of armored equipment. Although an unwieldy-looking machine, its low altitude attack capability and its armor made the aircraft a formidable opponent against Luftwaffe fighter planes. The early I1-2 had only two weak points which were vulnerable to its opponents; the oil cooler under the fuselage and the wooden empennage. With the arrival of the two-seat version of the I1-2, with its armored shield for the oil cooler, only one point of weakness was left, leaving beam fighter attacks from either side at low altitude most troublesome.

On 20 November 1942, the Russian Army opened a great counterattack against the German forces in the southern salient of the Stalingrad front. Poor weather made air support difficult, but the I1-2 pilots continued to make life miserable for the enemy. On November 21st, hero of the Soviet Union Captain V. M. Golubev, led his six I1-2s against a German airfield in the area. Despite heavy flak, the Russian fliers silenced the anti-aircraft guns and destroyed eight German aircraft on the ground.

On the return home the Russians encountered Messerschmitt 109's flown by pilots of the Hungarian Air Force who had joined in the Battle of Stalingrad. Defiantly, Golubev led his *Shturmoviki* into the enemy formation, with guns blazing. In the brief encounter the I1-2 pilots shot two of the Messerschmitts from the skies.

The *Shturmoviki* pilots never let up throughout the war. Their tank destruction efforts became legend and they proved to be the deciding factor in many tank battles. The "Black Death" more than lived up to its name and reputation.

Fiat C.R. 42 *Falco*

The success of designer Ing. Celestino Rosatelli's biplane, the C.R. 32, during the Spanish Civil War no doubt inspired the Italian *Regia Aeronautica* to accept his model C.R. 42 for production in 1936. The C.R. 42 was the last of the biplane fighters to be put into production by any major power.

The Fiat fighter was not fast by World War II standards, with a maximum speed of 266 miles per hour at just above 13,000 feet, nor was it heavily armed, possessing only a pair of 12.7mm machine guns firing through the propeller. The C.R. 42 did, however, maneuver well, had excellent flying qualities, and was a sturdy and durable aircraft.

Several foreign countries took interest in the *Falco* during its early production. The Hungarians took delivery of a number of the fighters in 1939, the same year that the Italian *Regia Aeronautica* began to assign the first operational models to various *Stormo*.

The first C.R. 42's to see combat were those assigned to the 2 *eme Groupe de Chasse* in Belgium. When, on 10 May 1940, that country was attacked by the Germans, the majority of the assigned C.R. 42's were caught on the ground where they were destroyed by bombing and strafing. Before their final destruction, however, a few of them did get airborne and entered combat with the Germans, and obtained three victories.

When Italy entered World War II on 10 June 1940, the Fiat C.R. 42 was the newest type of fighter assigned to the *Regia Aeronautica*. They were committed to action on 12 June 1940, when *Falcos* of the 23rd and 151st *Gruppos* attacked French airfields at Hyeres and Fayence where they destroyed some twenty aircraft on the ground and shot down one in aerial combat.

The C.R. 42 saw major action in both Italian East Africa and in the North African campaign. The *Falco* units were constantly in action against units of the Royal Air Force in both of these theaters as well as in the skies over Greece during the action in early 1941.

Classic amongst these actions were the last of the biplane dogfights. Depicted is a combat of renown that took place on 8 August 1940, between *Falcos* of the 9th and 10th *Gruppo*, 4th *Stormo* of the *Regia Aeronautica*, and No. 80 Squadron of the Royal Air Force which was

69

flying Gloster *Gladiators*. Nine of the C.R. 42's were shot down while the Italians claimed five of the *Gladiators*.

In the course of the action, Sgt. Lido Poli, who was seriously wounded early in the fight, continued the combat until he had downed one of the opposing *Gladiators*. Poli then made a forced landing and was immediately taken to the hospital, where doctors found it necessary to amputate a wounded arm. Sgt. Poli was awarded the Medaglio D'Oro for his courageous action.

The C.R. 42 continued to see action throughout the North African campaign, then followed the action as it progressed to Sicily and then Italy. When the Italian government surrendered in 1943, a number of *Falcos* were flown to Germany, where they saw limited action in combat but considerable duty as a training aircraft up to the end of the war in 1945.

German ingenuity outstripped all others to put the world's first jet racing in the sky, then the one and only rocket-powered plane to dive from high and flash through bomber formations to bewilder and confound the crews. V-1's and V-2's streaked across the Channel to rain down on southern England.

The names of Messerschmitt, Heinkel, and Junkers quickly became better known than those of twenty American presidents. Luftwaffe! *Blitzkrieg!* Schwarme and gaggle burst into the English language, to stay there even to this day.

A great air force, to be sure!

Here, then, are but twelve of these mighty air machines: Heinkels over London, wasp-nosed Bf 110's tearing up a Russian airfield, "Pip" Priller scattering *Lightnings*, the *Uhu* knocking down a *Mossie*, Bartels challenging the "Baby Air Force," and much, much more that will bring goosebumps to your flesh.

The paintings in this section are now a part of James Ross McDowell's private collection.

THE LUFTWAFFE GALLERY

Luftwaffe! Why, the very sight and sound of the word has the ring of audacity and the promise of a daring force to be reckoned with!

One can only wonder what would be our current state of aviation in general, and war-readiness in particular, had a high state-of-the-art not been present in the Luftwaffe when the hostilities began. The awesome power of the German war machine, the Luftwaffe in particular, forced the Allied aircraft designers to be extremely inventive, and to do it in a hurry.

Luftwaffe planes, pilots, strategies, and tactics forced a giant step forward in aerial warfare. The pace was greatly quickened. No longer could designers think in terms of biplanes or even the small and zippy monoplanes. Neither would the .303-caliber machine guns, usually just a pair of them, pose much of a threat to these powerful new German aircraft. The Germans came at you lightning fast, their cannons clearing the skies in short order; they were eagles among outdated birds of prey.

Fieseler Fi 156 *Storch*

The Fieseler Fi 156 *Storch* was probably the best all around observation and light reconnaissance aircraft of World War II. Its short takeoff and landing performance made it a superb aircraft for liaison and ambulance service.

When World War II began with the invasion of Poland in September of 1939, one of the first aircraft to fly reconnaissance missions ahead of the German Panzer forces was the Fieseler *Storch*. The aircraft saw yeoman duty during the *blitzkrieg* into France in 1940, and then went to North Africa with Rommel's legions. Its performance was legendary wherever it went and whatever the mission. Possession of the aircraft was a matter of position with German staff officers, and was a highly esteemed prize for Allied forces when they were able to capture one of the craft on the ground.

The flying weight of the *Storch* was only a little over 2,700 pounds, and its Argus eight-cylinder engine could propel the little craft along at a top speed of 165 m.p.h., but its normal cruise was only about 90 m.p.h. With a bit of head wind, and by utilizing its flaps, it could mush along at approximately 30 m.p.h., and, under certain conditions, it could become airborne in a scant 230 feet.

Its short takeoff and landing capability made the *Storch* a forerunner to the helicopter for transporting wounded from the battlefield. Many lives were saved by its ability to get into difficult terrain to perform these missions of mercy.

Following the invasion of Italy at Anzio on 22 January 1944, it was imperative for the U.S. Army Air Force to neutralize all Luftwaffe bases in the area. It was on just such a mission that the *Spitfires* of the 52nd Fighter Group encountered a small formation of German aircraft in the vicinity of Piombioa, Italy. Leading one squadron of the *Spitfires* was veteran Eagle Squadron pilot Lt. Richard L. "Dixie" Alexander.

Amongst the aircraft speeding away from the Luftwaffe field were two Fieseler *Storchs* towing little gliders. Alexander went down after one of the craft, whose pilot immediately cut the glider loose and then sped up to seek safety in a mountain pass. This highly maneuverable aircraft represented one of the hardest of all targets to hit, and "Dixie" Alexander knew it. He closed rapidly on the *Storch* and let fly with a burst while the German was in an evasive turn, and missed! Now "Dixie" pulled the nose of his *Spit* way 'round and gave the *Storch* a lot of deflection. The burst hit the little airplane in the tail section, and the *Storch* simply disintegrated.

Such occasions were very rare. Normally, fighter planes never got a chance to get a telling burst into the highly maneuverable craft. The few that were destroyed in combat were usually victims of ground fire.

To be sure, the many capabilities of the Fieseler *Storch* left an indelible mark in the annals of World War II.

Heinkel He 111

Hauptmann Albert Hufenreuter ignored the charts spread out in front of him. As navigator and aircraft commander of a Luftwaffe Heinkel He 111 medium bomber bound for England, he had flown over London, his target for tonight, many times before. Taking off from its base near Lille, in northern France, his He 111 was bearing northeast toward the English Channel. Soon the plane was lost in the murky black darkness of the night.

The date was 10 May 1941, and this was to be a maximum effort—"The final blow against London," according to the Nazi propaganda machine. All available aircraft of Luftwaffe West were to join in the attack on Britain's capital. Heading toward the chalk-white cliffs of Dover that delineated the English coast, were *Kampfgeschwadern* of Dorniers, twin-engined Bf 110's, and Heinkels, all flying at different assigned altitudes and with specifically allocated sectors to attack.

Prior to takeoff, at the briefing, the *staffel* leader had announced that following this mission the unit was being transferred to a new base, a great distance from Lille. He didn't say where—perhaps he didn't know yet himself. But this announcement served to add further apprehension to the mission.

Hauptmann Hufenreuter had mixed emotions about his aircraft. At times he loved the old Heinkel, and at other times he hated it. In recent missions she had brought him back safely from bombing runs against a British battleship under construction at Barrow-in-Furness, the Dumbarton shipyards in the Firth-of-Clyde, and, of course, from raids over London. She could cruise economically for up to nine hours at 300 kilometers per hour and was one of the safest planes Hufenreuter had ever flown. But in combat, it was another story. Her low speed and lack of maneuverability made the Heinkel easy prey for any fighter by day, and at night she was an even easier target for the night fighters.

The 111, which was powered by two 1,200 h.p. Junkers Jumo inverted V-type engines, had a range of 758 miles with a maximum bomb load of close to 2¼ tons. This range was far more than was needed for the short cross-channel run from northern France to England, and it provided a healthy reserve. The He 111 had been a civil transport before the war and required the addition of a vertical gondola in which a 20mm cannon was housed. A total of five 7.9mm machine guns mounted in the nose, dorsal, ventral, and beam positions completed the armament. It got its baptism of fire as a combat aircraft in the *Kondor Legion,* during the Spanish Civil War.

Bearing northeast, Hufenreuter ordered the sergeant-pilot to gain altitude as quickly as possible. Climbing, they made a wide swing, heading for the narrowest crossing of the Channel, between Calais and Cape Gris-Nez on the French side and Dover and Folkestone in England.

Because of the climbing, it took a half-hour to reach the British coast, at an altitude of about 6,000 meters. Dead ahead the crew could see London lit up by the hundreds of blazing fires. Dropping to about 2,000 meters, they could easily make out the big docks, sheds, factories, and other large buildings along the Thames River.

This night the sky was a melee of aircraft, probing searchlight beams, bursts of ack-ack, parachute flares, bomb flashes, the rising smoke, and British night fighters streaking in and out of the bomber formations.

Hufenreuter released his bombs on target and while still over London, turned south toward the Channel and climbed steeply to 4,000 meters. Leaving that raging hell behind him, Hufenreuter warned his four-man crew, "Keep your eyes open, boys! We aren't out of it yet! Watch out for night fighters!"

His warning was not without substance. Within a matter of seconds he saw a burst of about a dozen tracer bullets whip by below the left wing. "Night fighters! Dive, dive!" he shouted. But it was too late. A second burst caught the port engine.

Hufenreuter never knew what hit his aircraft. It was probably a *Hurricane,* but he didn't see it. The Heinkel dived, took evasive action, and appeared to have lost the attacking fighter.

The instrument needles had gone haywire, and at 1,500 meters the port engine stopped completely. At an altitude of 1,000 meters they could see the silvery water of the Channel up ahead. But the trees and bushes were rushing up fast as the pilot shouted, *"Hauptmann,* I can't hold her . . . !"

Hauptmann Hufenreuter and his crew sat out the rest of the war in a British P.O.W. camp. Following V-E Day, he returned to Germany, where he became an English teacher.

Messerschmitt Bf 109-G *Gustav*

During the course of World War II, the Luftwaffe constantly modified and up-dated its first line fighter—the Messerschmitt Bf 109.

Upon the arrival of the *Spitfire* Mark V as an opponent, the Messerschmitt Bf 109-E gave way to the Bf 109-F, with its rounded wing tips, modified fuselage, and new engine. During the summer of 1943,

the fighter units of the Luftwaffe began to take delivery of what was perhaps the most famous of the Messerschmitt Bf 109 line—the "G", or *Gustav*—most of which incorporated a new armament, and all of which were powered by a 1,475 horsepower Daimler-Benz engine. The new armament package consisted of the customary engine-mounted 20mm cannon firing through the propeller hub, but instead of the two 7.9mm machine guns firing through the propeller, it mounted two 13mm machine guns, which necessitated fairings on the gun breeches that gave the model its noticeable bulges just forward of the cockpit.

By late 1943 the *Gustav* was standard equipment for most of the *Geschwader* in the Luftwaffe. Amongst those based in the Mediterranean who were flying the craft was the IV/JG 27 at Kalamaki in Greece. A young German pilot, who wore the Knight's Cross, was assigned to the unit that summer. He was *Oberfeldwebel* Heinrich Bartels, and he had won the coveted Knight's Cross for downing 47 Russian aircraft on the Eastern Front.

Opposite the Luftwaffe forces in Greece was the "Baby Air Force" of the Mediterranean, which was composed of the B-25 medium bombers of the 12th Air Force's 321st Group, supported by the veteran P-38-equipped 82nd Fighter Group. The bombers were assigned targets in Greece, Yugoslavia, and Albania, as a rule, while the P-38's escorted them, giving battle to the Luftwaffe fighters and doing bombing and strafing work of their own.

On 15 November 1943, the *Mitchells* of the 321st Group set out to destroy the airbase at Kalamaki which was proving to be a real thorn in the side of the "Baby Air Force." As usual they were escorted by the P-38's of the 82nd Fighter Group. Over the target, they were attacked in force by the Messerschmitt *Gustavs* of IV/JG 27. In the heated air battle, *Obw.* Bartels downed one of the *Mitchell* bombers as well as one of the *Lightnings.* In the constant action of late 1943, Bartels would score some 20 victories while flying against the "Baby Air Force" and other units of the A.A.F. and R.A.F. operating in the Aegean area.

The pilots of the Luftwaffe continued to operate the Bf-G's until the end of the war. Many of the higher scoring aces of the Luftwaffe on the Eastern Front scored the majority of their successes flying the famed *Gustav.* Bartels would go on to score 99 victories in the craft before he was killed in action on 23 December 1944 while flying on the Western Front.

Henschel Hs 129

The Henschel Hs 129 had to overcome an "ugly duckling" reputation before it won fame in the Luftwaffe. Although it was originally designed as a ground support aircraft, the Hs 129 was far from an immediate success. In its original conformation the aircraft was underpowered, the pilot's view was restricted considerably, and flight performance was very poor.

The Luftwaffe refused to utilize facilities in Germany for modification, so the Hs 129 was refitted with captured French Gnome Rhone radial engines. This improved performance to some extent. The armament combinations carried by the aircraft were formidable, with two 7.9mm machine guns and two 20mm cannons for basics.

The aircraft saw its initial action in North Africa, where the Gnome Rhone engine was no match for the desert dust. After numerous engine failures the Hs 129's were pulled out of combat and sent to the rear for refitting with sand filters, but to no great avail.

In late 1942 the Hs 129's were sent to Russia, where they were to perform anti-tank duty, and it was here that the little twin-engined craft won its fame. During July of 1943, the Henschel Hs 129 wrought tremendous havoc upon Soviet tanks, and on one occasion repulsed an entire armored brigade. Operating in relays, four squadrons of Hs 129's destroyed most of the tanks during the concentrated attack.

As the war progressed, however, the Soviets developed new tanks which could not be stopped with the 20mm or, later, 30mm cannons that were used by the pilots of IV (Panzer) *Gruppe*/SG 9. It was found through experimentation that a 75mm gun could be used effectively when fired from the Hs 129. This electropneumatically operated gun was fitted on the aircraft and sent into action against the giant "Josef Stalin" tanks during the winter of 1944–45.

In January of 1945, one of the most outstanding tank destroyers of the Luftwaffe became Group Commander of I(PZ) SG 9. This was *Hauptmann* Andreas Kuffner, who had destroyed more than 50 tanks while flying the Hs 129. Kuffner continued his success and was the victor several times while using his 75mm cannon against the heavily armored "Josef Stalin" tank. Before his death in April of 1945, Kuffner had bagged in excess of 60 tanks.

The "ugly duckling" of the Luftwaffe had become the Panzer scourge of the Russian tank forces and left a record that is unparalleled in anti-tank warfare right to this day.

Focke Wulf Fw 190

Although it was not produced in the quantity of the Messerschmitt Bf 109, the Focke Wulf 190 absorbed the brunt of the action on the Western Front from early 1942 until the end of the war. Its big radial engine, clean lines, and heavy armament made the airplane an excellent choice to be operated in a number of roles. However, it was as an interceptor of the Allied bomber streams that the 190 gained its fame.

Beginning in the spring of 1942, the Focke Wulf 190 saw yeoman duty in the skies over France and the Channel coast, where it engaged in combat with the bombers and intruders of the Royal Air Force. When the American daylight bombardment campaign began that summer, it was the Focke Wulf 190 that made the interceptions and all but made the plan fail due to unacceptable losses to the bomber forces.

Once the heavy bombardment really got underway during daylight hours, the Focke Wulf 190's of JG 1 and JG 26 were fitted with rocket tubes to add to the existing firepower from the two machine guns and four 20mm cannons which were standard armament. On "Black Thursday," 14 October 1943, when the *Fortresses* went deep into Germany, to Schweinfurt, which was beyond the range of any fighter escorts at that time, it was the Focke Wulf 190's of these two units that rose to do battle and downed the majority of the sixty-two B-17's that failed to return from the mission.

However, the ever-increasing number of Allied fighters, and the advent of the P-51 *Mustang* as an escort fighter capable of going with the bombers to any target in Germany, began to take its toll on the small number of Luftwaffe fighters available in the West. Coupled with a heavy loss of experienced pilots in the East and the Mediterranean, the German Fighter Force did well to mount the attacks they did throughout the great air battles in the spring of 1944.

When D-Day in Normandy came on 6 June 1944, most of the Focke Wulfs had just moved from their bases on the Channel coast to bases further south to escape the constant pounding they were taking from both Allied bombers and fighters. So it happened that on D-Day only the *Kommodore* of JG 26, Major Josef "Pip" Priller, and his wingman were available to face the monstrous invasion. Undaunted, Priller and *FW.* Wodarczyk made a strafing run along the beach and then speedily and prudently left the scene.

Five days later Priller scored his 99th victory. The P-38 *Lightnings* of the 55th Fighter Group, under the command of Major Giller, were assigned to attack the German airfields and marshalling yards south of the invasion beaches. When the alarm came, Priller led his squadron of Focke Wulf 190's into the air to do battle with the Americans. Contact was made in the vicinity of Ressons, where a heated air battle immediately took place between ten Focke Wulf 190's of JG 26 and ten to twelve P-38's of the 55th Fighter Group.

Priller dived down and immediately slid in on the tail of one of the *Lightnings*. A burst from his cannon set the aircraft aflame and it went down. Lt. Gerd Wiegand scored telling hits on another one of the *Lightnings,* then watched as the pilot bailed out when the craft's engine burst into flames. Two of the P-38's were downed and several others suffered heavy damage, while three of the Fw 190's failed to return to base.

Four days later Priller scored his 100th victory, for which he was awarded the Oak Leaves with Swords to the Knight's Cross. This great German fighter pilot was to score only once more: a P-51 *Mustang* on 12 October 1944. Priller's 101 victories were all scored on the Western Front and, miraculously, he was never shot down in all his years of air combat.

The Focke Wulf 190 continued to see constant combat right up to the end of the war. Its later version, which incorporated the Junkers Jumo in-line engine, was one of the finest fighter planes to see action during the World War II era. The long-nosed 190 took a heavy toll of Allied bombers and fighters alike before the hostilities ceased in May of 1945.

Messerschmitt Me 410

Throughout World War II the Germans attempted to come up with a real destroyer-type aircraft that could wreak heavy destruction on the American bomber formations. The Messerschmitt Me 210 was one of the aircraft built specifically to deal with the problem. As a bomber-destroyer, it possessed heavy armament in the nose section, with two 20mm cannons and two 7.9mm machine guns. The plane also carried two remote-controlled rearward-firing 13mm machine guns that were mounted on barbettes attached to the sides of the fuselage. The aircraft's speed of 385 m.p.h. was good and the approximately 1,500-mile range was satisfactory. But, alas, it was loaded with "bugs." The plane suffered from all of the different ailments known to plague combat-type aircraft, and its accident rate was appalling. In short, the 210 was a dismal failure.

Then, in 1943, its successor made its appearance. This was the Messerschmitt Me 410, which was basically the Me 210 with a new engine. The bug-riddled plane's performance improved immediately with installation of the new engine, and in late 1943 and early 1944 the Me 410 began operating successfully as a fighter-bomber over the British Isles, where it became a formidable opponent to the R.A.F.'s *Mosquito* night fighter.

The Luftwaffe, however, still needed a good bomber-destroyer to carry a heavy weapon that could take to task the American daylight bombers, yet could remain out of range of the .50-caliber machine guns of the *Fortresses* and *Liberators*. In early 1944, German technicians began experimenting with the use of the BK 50mm cannon in the nose section of the Messerschmitt Me 410. This installation proved to be so successful that a number of Me 410A-2's on the production line were modified to utilize the weapon.

Accordingly, the cannon-armed Me 410 was supplied in good numbers to II/ZG 26 for home defense against the heavy bomber formations. Initially, the unit scored numerous successes with the heavy weapon. Approaching the bomber formation from abeam, the Me 410 pilots would commence blasting away with the deadly cannon while standing well out of range of the bomber's guns. Losses were few, while damage to the helpless bombers was great.

II/ZG 26's luck ran out, however, on 13 May 1944. On this day B-17's of the 8th Air Force's 96th Bombardment Group came under a savage attack by the Me 410's until P-51's of the 335th Fighter Group arrived on the scene. The *Mustangs* raced into the sluggish Me 410's and broke up the attack immediately, blasting six of the 410's from the sky in quick order. As the Messerschmitts fled from the melee they were jumped again, this time by P-51's of the 357th Fighter Group, which further decimated the ranks. II/ZG 26 suffered such heavy casualties that they never again took their 410's into action against the bombers.

The Messerschmitt Me 410 did, however, continue to operate in the various theaters of Europe until the end of the war, but never again did it enjoy any degree of success as a destroyer of American bomber formations.

Messerschmitt Me 262

Throughout the early years of World War II, Adolf Hitler boasted of the German "secret weapons" that would change the complexion of warfare. How ironic it was that his decree to make the Messerschmitt Me 262—one of his vaunted "secret weapons"—a bomber, rather than a fighter, probably prevented the Allies from losing air superiority over Europe!

The Me 262 first flew on 18 July 1942, and while not the first German jet to fly, it was by far the most superior. Many mechanical and engineering problems held up early production of the Junkers Jumo jet-propelled aircraft, but by late 1943 four prototypes had been built and it was decided the time had come to demonstrate the Me 262 to Hitler, in the hope that he would order it into mass production.

When the jet flew for Hitler at Insterburg on 26 November 1943, he was greatly impressed, but he acclaimed the aircraft to be a high speed bomber. From that day forward any production of the Me 262 that was directed toward fighter operations had to be handled very discreetly.

The Me 262 went operational with a test group in September of 1944,

and these crafts were transferred to *Kommando Nowotny,* which was led by the high scoring Luftwaffe ace, Major Walter Nowotny. This group met with limited success in combat and was disbanded after the loss of their leader on 8 November 1944.

As a bomber, the Me 262 did go into service in the fall of 1944, but, due to its speed and the limited bomb load, its missions could only be described as nuisance raids.

Fighter Forces organized the jet-propelled force into *Jagdgeschwader 7 Nowotny* in late 1944 and the unit scored its first victory on 26 December 1944. With experienced pilots at the controls, the four 30mm cannon-armed crafts were quite effective in high speed passes against Allied bombers, but it was felt that they could be made much more deadly if rockets were also used. After considerable experimentation, a number of Me 262's began utilizing as many as twenty-four 55mm R4M rockets, which were mounted on rails under the wings.

JG 7 met with considerable success despite great odds and operational difficulties. The Allies had no fighter plane that was capable of catching the Me 262, and the German pilots could avoid combat with the bomber escorts at will. The Allies, however, held air superiority and their fighters chose to orbit the Luftwaffe jet bases where they could attack the Me 262's during takeoffs and landings. Even when the Luftwaffe used Messerschmitt Bf 109's and Focke Wulf Fw 190's to protect the bases, the strategy didn't prove to be too effective in deterring the tenacious Allied fighters. Once in the air, however, the Me 262 performed brilliantly. By the end of the war, JG 7 alone had claimed some 400 victories, with 300 of them being four-engined Allied bombers.

In January of 1945, the most celebrated fighter unit of WW II was formed. General Adolf Galland, Commander of the German Fighter Forces, had fallen from grace with the Nazi High Command and was relegated to combat duty. Galland formed *Jagdverband* (JV) 44, an elite unit of veteran fighter pilots. Of the 50 pilots assigned to JV 44, ten were holders of the coveted Knight's Cross.

JV 44 operated in flights of three aircraft during combat. Upon sighting a formation of Allied bombers they would form-up high and to the rear. As they dove down in a high-speed pass, they would first fire their rockets and then follow up with 30mm cannon fire. Once the pass had been completed, they would go into a high-speed flat climbing turn just over the bomber formation. At this speed it was most difficult for the gunners in the bombers to track rapidly enough to hit them.

One of the more actively concentrated days of attack by the Me 262's against the Allied bomber formations was 7 April 1945. Fifteen American heavy bombers were lost that day, most of them to the German jets. One of the B-24 formations attacked was that of the 8th Air Force's 734th Squadron of the 453rd Bomb Group. As the gunners blazed away, the Me 262's of JV 44 dived down in a defiant pass. Two of the bombers fell away in flames, but one of the attackers went down trailing heavy smoke. The jets could be hit!

But for Hitler's order to make it be a bomber, many more American bombers would have fallen to the rockets and guns of the world's first jet-propelled fighter.

Focke Wulf Fw 200 *Condor*

"The Focke Wulf and other bombers employed against our shipping must be attacked in the air and in their nests!" was the order Winston Churchill gave to his fighting forces on 6 March 1941. It was, however, too late to save the *Empress of Britain*.

On 27 May 1931, one of the most magnificent ships of her day slid down the ways at Southampton. The Prince of Wales—later, King Edward VIII—did the honors by smashing a bottle of champagne against the bow of the second *Empress of Britain*. On her maiden voyage, she carried Mary Pickford and Douglas Fairbanks among her distinguished passengers. She was the largest and most luxurious vessel put into service by the Canadian Pacific Line up to that time. The 42,000 ton *Empress* was the first ocean liner equipped with ship-to-shore radio-telephone. Among her many other features were a full-size doubles tennis court, squash court, indoor Olympic swimming pool as well as a smaller open deck pool, a fully equipped gym, Turkish baths, and more. Like the *QE II* and *Rotterdam* of today, she sailed on round-the-world cruises. The *Empress* was truly a magnificent ship.

In July 1937, Kurt Tank, who was shortly to design the great German fighter plane—the Fw 190—sat at the controls of his new 26-passenger airliner on her first flight. He was pleased with the results of his engineering skill. The four-engined Focke Wulf 200 *Condor* also performed spectacularly in flights from Berlin to New York and return, and from Berlin to Tokyo.

Like the Dornier Do 17 and the Heinkel He 111, the *Condor* originally entered service as a civilian transport plane for *Deutsche Lufthansa*. Designing civilian aircraft, with the military mission in mind, was a ruse employed by the Nazis to get around the Treaty of Versailles that prohibited Germany from re-arming. The treaty, which had ended World War I, sought to prevent Germany from ever again becoming a threat. Obviously, it failed.

In 1939, the *Empress* and the *Condor* received new assignments. They both went to war. The ocean liner was converted to a troopship and the airliner became a bomber.

In Germany, Luftwaffe *Oberst Leutnant* (Lt. Colonel) Edgar Peterson was ordered to select a suitable long-range maritime patrol bomber and to organize a *staffel*. He picked the *Condor*.

The Fw 200 first saw action carrying supplies to an isolated garrison during the Norwegian campaign. Its principal mission, however, was to seek out and destroy British shipping. Organized on 1 October 1939, Peterson's outfit was designated the First *Staffel* of *Kampfgeschwader* 40, or KG 40, early in 1940. By July of that year the *staffel* was increased to the strength of a group, and later another group was added. All of KG 40's *Condors* bore the unit's "World-in-a-Ring" insignia slightly aft of the pilot's compartment on both sides, and each had the name of a different star or planet across its nose.

On 26 October 1940, the *Empress of Britain* met a Fw 200 *Condor* seventy miles off the northwest coast of Ireland. *Hauptmann* (Captain) Bernhard Jope was on his first mission with KG 40 when he spotted the troopship. The *Empress* had few, if any, air defenses, and Jope brought his Fw 200 in very low over the water. The *Condor's* bomb sight was not effective at low altitude, but the bomb release was rigged to drop the five 550-pound bombs at eight yard intervals. Two of them struck home, leaving the once proud *Empress* crippled and on fire. Two days later, while being towed by a Polish destroyer, she was torpedoed and sunk by the German submarine U-32.

The tactics employed by Jope were used effectively by the rest of the *geschwader,* and during the first quarter of 1941 the *Condors* sank 88 ships for a total of 390,000 tons. This was accomplished by no more than eight Fw 200's, the maximum KG 40 could muster at any one time.

No fools the British, they soon armed their merchant ships and the picnic was over.

Heinkel He 219 *Uhu*

Only two aircraft that had been designed from the ground up as night fighters saw action during World War II. These were the Heinkel 219 *Uhu* (Owl) of the Luftwaffe and Northrop's P-61 *Black Widow* of the U.S.A.A.F. The excellent Bf 110 *Zerstörer* was, like the R.A.F.'s *Mosquito,* an adaptation of an existing design and not a purposely built night fighter.

Design work on both of these night fighters began in 1940 and, as it turned out, each were to be relatively large, twin-engined machines. Production of the *Uhu* was slow and limited, hence the number on hand at any given time was not very large. This was due to the fact that day fighters and more established aircraft were given priority position on the assembly line. Although the *Uhu* was much heavier than the *Black Widow,* it achieved a higher top speed, greater altitude, longer range, and carried a heavier armament package on slightly less powerful engines. The *Black Widow* did, however, possess the better radar equipment. On the other hand, *Uhu* possessed a very special innovation—it was the first major service type to feature ejection seats for its crew.

The Heinkel claimed its first victories a full year before the *Black Widow.* Neither plane was used on a wide-scale basis until in the latter part of 1944. And the German night fighter emerged as the better performer of the two.

Carrying both forward-firing guns and the upward-firing *Schräge Musik* (the literal translation is "slanting music," but is, in fact, the German term for "Jazz") installation, the He 219 was capable of inflicting untold havoc on a bomber stream. During the first operational sortie with a pre-production test aircraft, the Luftwaffe's leading night fighter pilot of the period, Major Werner Streib, claimed five bombers shot down during the night of 11/12 June 1943. During the first six sorties made by He 219's in the summer of 1943, twenty British aircraft fell victim to the *Uhu's* guns, including no less than six *Mosquitos.* Indeed, the aircraft proved to be the only German night fighter in service prior to 1945 that had any real chance of catching or combatting the various versions of the formidable *Mosquito,* which at the time was operating virtually unmolested over Germany as a bomber, photo-reconnaissance aircraft, and intruder fighter, and increasingly as a night counter-air and escort fighter.

Our scene depicts a He 219 *Uhu* of 1/*Nachtjagdgeschwader* 1, the main unit employing these aircraft, scorching a *Mossie,* its most difficult quarry.

Messerschmitt Me 163 *Komet*

Although 8th Air Force commanders had known for some time that the German Luftwaffe was readying jet- and rocket-powered aircraft, it was not until July of 1944 that this menace made its formal appearance in the skies. Colonel Avelin Tacon, C.O. of the 359th Fighter Group, reported sighting, and attempting to combat, two of the rocket-powered Messerschmitt Me 163's on 28 July 1944.

All 8th Air Force units took notice of the report and readied themselves for possible attacks. The following day, Captain Arthur Jeffrey, of the 479th Fighter Group, encountered one of the speedy craft and managed to put some shots into it before he lost the *Komet* in a 500 m.p.h. dive.

By August of 1944, I/JG 400, the unit equipped with the Me 163, was based at Brandis, near Leipzig, Germany. It was hoped that from here the interceptors would be effective in driving the American bombers from the industrial plants in the area.

When on 16 August 1944 the heavy bombers of the 8th Air Force attacked targets in this area, the Luftwaffe intercepted them in strength. A B-17 *Flying Fortress* named "Outhouse Mouse" from the 91st Bombardment Group was one of the bombers to encounter the full force of the German attackers. First, "Outhouse Mouse" was knocked out of formation by a Fw 190 and a Messerschmitt 109, both of which inflicted material damage to the bomber and wounded two of its crew members.

Then, as it limped along, it came under attack from a flight of two of the Messerschmitt Me 163 *Komets*. The German pilot brought his rocketing fighter down in a screaming dive, aiming on the tail of "Outhouse Mouse," while Lt. Reese Mullins took every kind of evasive action he knew in the attempt to escape the attacker's cannon fire. And Mullins was apparently effective, for the Me 163 pilot couldn't seem to get his guns lined up on the B-17.

Simultaneous with the German's attack, Lt. Colonel John B. Murphy and Lt. Cyril W. Jones of the 359th Fighter Group spotted the Me 163's going after the crippled *Fortress* and came roaring down to give assistance. There was no way they could catch the German in his dive, but after the *Komet* came out of the dive, Murphy caught him, got in a short burst, and then overran the craft. Lt. Jones caught the Me 163 as it rolled over and put a good burst into it, but then blacked out when he tried to follow it through on its dive.

Murphy then caught another Me 163 making shallow diving turns at a lower altitude. It could have been that the 8 to 12 minute fuel supply had been expended by the German pilot. Regardless, Murphy rapidly overtook him and scored hits on the fighter until a violent explosion occurred; the disintegrating *Komet* continued its downward plunge, streaming fire and parts all the way to the ground.

Although it attracted a lot of attention, the Messerschmitt Me 163 was never a combat success. Very few Allied aircraft were downed by it, while many of the German pilots fell victim to the erratic performance of their own aircraft.

Junkers Ju 88

The Junkers Ju 88 was probably the most versatile twin-engined aircraft utilized by the Luftwaffe in World War II. It was a medium altitude bomber, a dive bomber, a daylight fighter against the Allied bomber streams, and a long-range night intruder or interceptor, as the case might be.

The initial unit to take the Junkers Ju 88 into combat was KG 30, which struck the British Fleet in the Firth-of-Forth on 26 September 1939. KG 30 served with distinction throughout the Battle of France, the Battle of Britain, and took part in the initial onslaught on Russia.

Spring of 1942 saw the Allies trying desperately to get supplies to Russia across the North Atlantic. Slowly, but surely, progress was being made in the anti-submarine campaign, and, as the result, the German U-boat concentrations were being pushed further and further eastward. The wolf packs of U-boats and long-range German aircraft were, however, still taking their toll of the merchant vessels steaming in convoys bound for the Russian port of Murmansk.

One of the Luftwaffe units that was striking against the convoys from its base in Norway was III *Gruppe* of the famed KG 30. Among the veteran pilots assigned to III/KG 30 at that time was Major Werner Baumbach, who would win the Knight's Cross with Swords for his efforts against the convoys. Baumbach had seen extensive action since the beginning of the war and both he and his crew knew no peers when it came to anti-shipping strikes.

On a sunny Sunday afternoon in the spring of 1942, Baumbach lifted his Ju 88 from the runway of the airfield at Stavanger-Sola in Norway and climbed steadily as he took a westward course. Long-range reconnaissance aircraft had reported Allied supply vessels were docked in the Danish-owned Faeroe Islands, which lay in the North Atlantic between Scotland's Shetland Islands and Iceland. Baumbach and his Ju 88 crew had been waiting for a break in the weather to strike at these vessels, and today was the day.

The course to the Faeroe Islands took the Ju 88 far out to sea, nearly 375 miles from their home base. After an uneventful flight, the jagged, rocky summits of the islands made their appearance above the clouds. Baumbach maneuvered the Junkers around to come in the target area from the west. The clouds seemed to part in the target area, and the Junkers came roaring in over Thornshavn fjord. There before Baumbach was a large merchant vessel, with several smaller ships behind it, sitting out in open waters.

Baumbach nosed the Junkers over in a diving attack and swiftly closed the distance to the target. Quickly, he lined up on the big ship as its decks grew larger and larger in his sights. Then, a push of the button and the bombs were away. The first projectile fell short, but the second scored a direct hit on the port side of the ship. Black clouds of smoke rose from the target as Baumbach set the Junkers on a course for the long flight home.

The following day a reconnaissance plane reported that the ship was burned out and listing badly. Major Werner Baumbach had added another victim to his growing list of Allied ships sent to the bottom. Sixteen ships would feel the wrath of his strikes before he was promoted to command the German bomber forces for the balance of World War II.

Messerschmitt Bf 110 *Zerstörer*

The new Bf 110 *Zerstörer* (destroyer) was just entering service when the outbreak of World War II came in September of 1939. The aircraft had fired the enthusiasm of the Luftwaffe Commander-in-Chief, Hermann Göering, and the units flying *Zerstörers* were the "apple of his eye."

The *Zerstörergruppen* achieved notable successes when faced with low-performance adversaries, or unescorted bombers, over Poland, France, and the Low Countries, and over the North Sea and Norway. But when challenged by *Hurricanes* and *Spits* over Dunkirk, *Zerstörer* pilots suffered their first major setback, and the losses were heavy: The plane was unable to maneuver effectively against these British fighters. Moreover, the *Zerstörer* proved to be a dismal failure as a long-range escort for the Luftwaffe's bombers. In this role their losses bordered on being catastrophic; hence, the bombers, for all practical purposes, went unescorted. In the end, the *Zerstörers* had to have their own escorts of Bf 109's!

Events during the winter of '41, however, caused the Bf 110 to be tried out in the role of a night fighter. This was brought about by the ever increasing night raids by British bombers. Because the aircraft was a heavily armed, stable gun platform, the Bf 110 was an excellent bomber-destroyer and, as such, racked up many successes at night. As a result, the aircraft remained one of the numerically important types in the Luftwaffe's night defense force for the remainder of the war.

Campaigns in the Balkans and the North African Desert provided the means to give another role, and a new chance, to the *Zerstörergruppen* still operating by day. They were employed down on the deck for ground attack missions, and the results were good, particularly when they went after an airfield. These successes took four *Gruppen* of Bf 110's to Russia when that country was invaded in 1941, and it was on this front where the *Zerstörer* amassed a legion of successes both as a day fighter and destroyer of ground targets. This was due primarily to the initial low quality of the opposition. In time this shortcoming of the Russians would change, but for the moment the *Gruppens* ran amuck, destroying hundreds and hundreds of Russian aircraft both in the sky and on the ground, laying waste to great numbers of tanks, trucks, gun emplacements, rail locomotives and rolling stock, and other such targets.

Our scene shows Bf 110's of II/ZG 1, the *Wespengruppe* (Wasp group), attacking Russian fighters caught napping on the ground in southern Russia.

All *Zestörergruppen* were subsequently withdrawn from Russia and returned to Germany, once again to go up as bomber destroyers to help counter the growing numbers of the 8th Air Force's B-17's and B-24's appearing over the *Reich*. Initially, the heavily armed and rugged Bf 110 proved extremely efficient at this task. Their destructive powers were short-lived, however, for with the appearance of long-range fighter escorts, coming with the bombers to targets deep inside Germany, the Bf 110's suffered terrible losses and were quickly driven from the skies.

COLOR PLATES
Volumes I and II

Curtiss P-40 *Warhawk (Tomahawk)*

Hawker *Hurricane*

Supermarine *Spitfire*

Republic P-47 *Thunderbolt (Razorback)*

Messerschmitt Bf 109

North American P-51 *Mustang*

Yakovlev *Yak*

Republic P-47 *Thunderbolt*

Lockheed P-38 *Lightning*

Bell P-39 and P-400 *Airacobra*

Grumman F4F *Wildcat*

Grumman F6F *Hellcat*

Mitsubishi A6M5 *Zero*

Chance-Vought F4U *Corsair*

Grumman F8F *Bearcat*

Nakajima Ki-43 *Oscar (Kamikaze)*

Boeing B-17 *Flying Fortress*

Boeing B-29 *Superfortress*

Consolidated B-24 *Liberator*

Martin B-26 *Marauder*

North American B-25 *Mitchell*

Dornier Do17

Junkers Ju 87 *Stuka*

Douglas A-20 *Havoc (Boston)*

Douglas SBD *Dauntless*

Douglas A-26 *Invader*

Curtiss SB2C *Helldiver*

Grumman TBF *Avenger*

T. Weddel

De Havilland *Mosquito*

Aichi Type 99 D3A1 *(Val)*

Lockheed *Hudson*

Consolidated PBY *Catalina*

Vought-Sikorsky OS2U *Kingfisher*

Stinson L-5 *Sentinel*

Messerschmitt Bf 108b *Taifun*

Missing Man Formation

Westland *Lysander*

Vickers *Wellington*

Boulton-Paul *Defiant*

Fairey *Swordfish*

Supermarine *Spitfire*

Gloster *Gladiator*

Bristol *Beaufighter*

Avro *Lancaster*

Handley-Page *Halifax*

Hawker *Tempest*

Hawker *Typhoon*

Gloster *Meteor*

Northrop P-61 *Black Widow*

Brewster F2A-3 *Buffalo*

Boeing P-26A *Peashooter*

Bell P-63 *Kingcobra*

P.Z.L. P-11

Douglas TBD *Devastator*

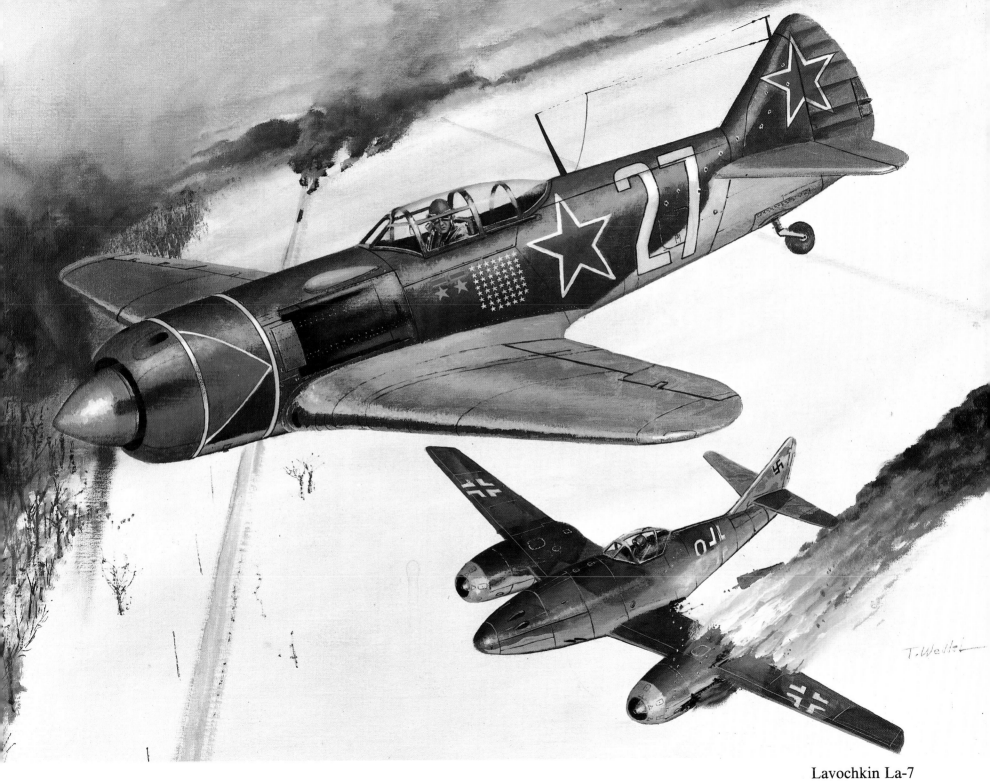

Lavochkin La-7

North American P-51-B *Mustang*

Macchi C. 202 *Folgore*

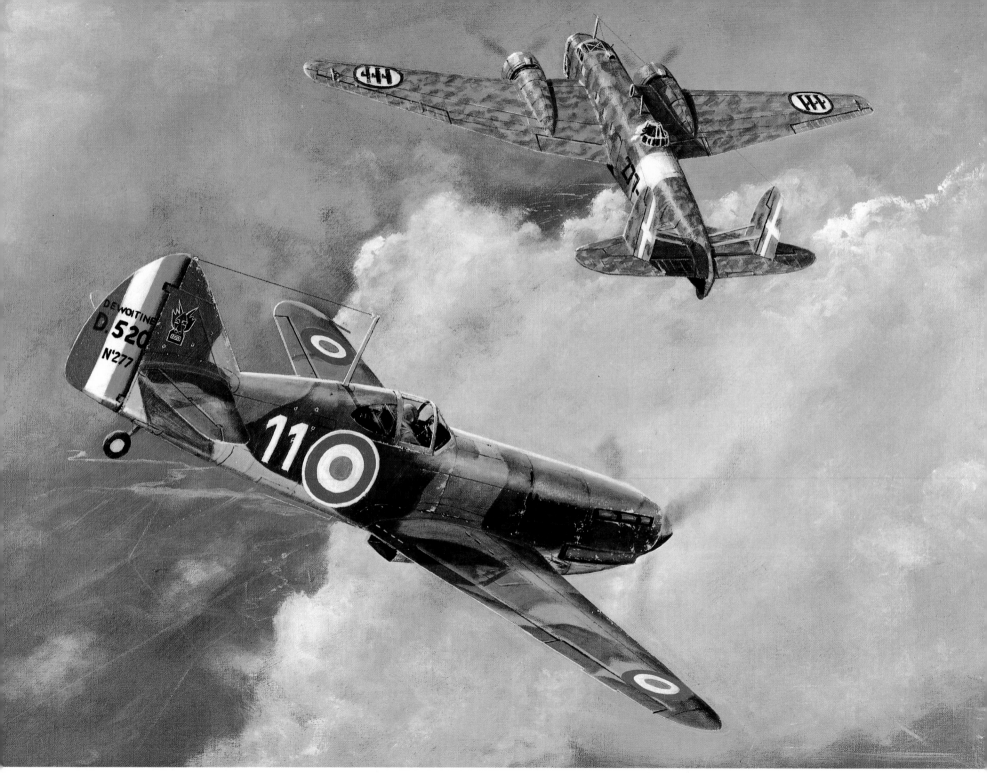

Dewoitine D. 520

Ilyushin Il-2 *Shturmoviki*

Fiat C.R. 42 *Falco*

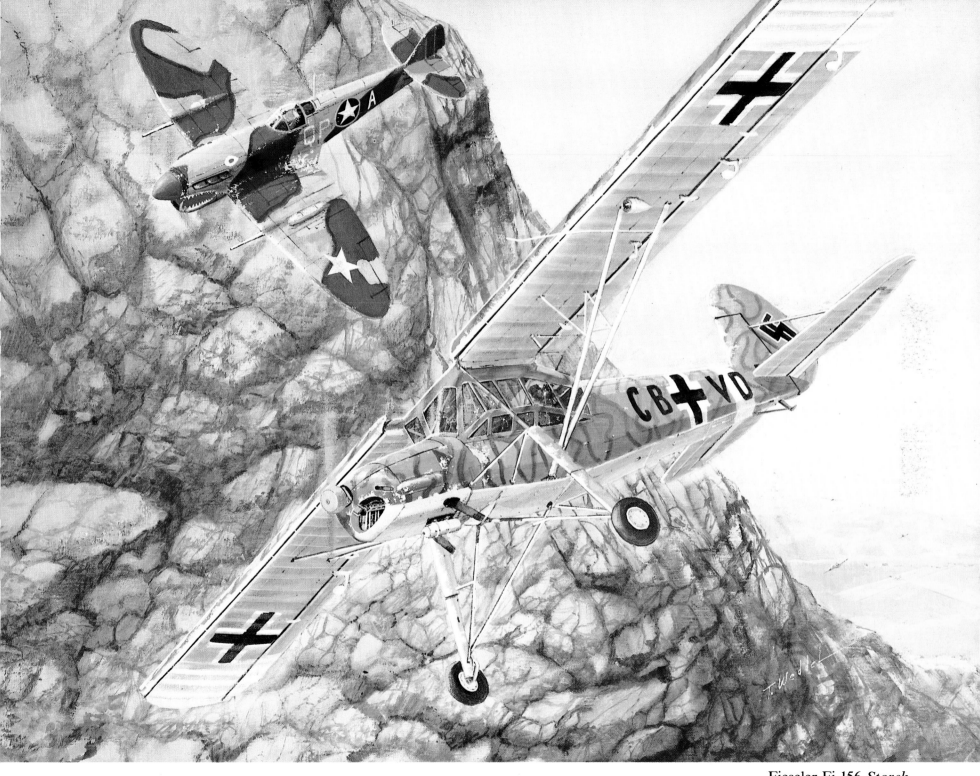

Fieseler Fi 156 *Storch*

Heinkel He 111

Messerschmitt Bf 109-G *Gustav*

Henschel Hs 129

Focke Wulf Fw 190

Messerschmitt Me 410

Messerschmitt Me 262

Focke Wulf Fw 200 *Condor*

Heinkel He 219 *Uhu*

Messerschmitt Me 163 *Komet*

Junkers Ju 88

Messerschmitt Bf 110 *Zerstörer*